★

TURKEY'S NUCLEAR FUTURE

EDITORS

GEORGE PERKOVICH
SINAN ÜLGEN

★

TURKEY'S
NUCLEAR
FUTURE

CARNEGIE
ENDOWMENT FOR
INTERNATIONAL PEACE

Carnegie Endowment for International Peace
1779 Massachusetts Avenue NW
Washington, DC 20036
P+ 202 483 7600
F+ 202 483 1840
CarnegieEndowment.org

The Carnegie Endowment does not take institutional positions on public policy issues; the views represented here are the authors' own and do not necessarily reflect the views of Carnegie, its staff, or its trustees.

This research has been funded in part with support from the William and Flora Hewlett Foundation.

The opinions expressed in this report are those of the authors and do not necessarily reflect the views of the William and Flora Hewlett Foundation.

To order, contact:
Hopkins Fulfillment Service
P.O. Box 50370
Baltimore, MD 21211-4370
P+ 1 800 537 5487 or 1 410 516 6956
F+ 1 410 516 6998

Cover design by Jocelyn Soly
Composition by Cutting Edge Design
Printed by United Book Press

Library of Congress Cataloging-in-Publication Data for this volume are available

ISBN 978-0-87003-415-2 (pbk.) -- ISBN 978-0-87003-416-9 (cloth) -- ISBN 978-0-87003-417-6 (electronic)

MIX
Paper from
responsible sources
FSC
www.fsc.org **FSC® C010236**

CONTENTS

FOREWORD

The international landscape is undergoing significant change—new global and regional powers are rising, hundreds of millions of people around the world are climbing into the middle class, hyper-empowered individuals with the capacity to do great good and huge harm are multiplying, and more information is flowing more rapidly than ever before. Turkey's emergence over the past decade as a more independent and assertive foreign policy actor is indicative of this change and both the challenges and opportunities it presents to regional stability and international security.

This volume focuses on one of the most consequential aspects of Turkey's transformation—its quest to enhance its energy security through nuclear power. *Turkey's Nuclear Future* provides a comprehensive and unique examination of the different yet interdependent dynamics of nuclear policymaking in Turkey. The volume also covers new ground by providing an insightful analysis of the country's emerging role in the governance of the nonproliferation regime. Finally, the volume provides a detailed examination of how Turkey might react to Iran's acquisition of nuclear weapons and how Turkey and key international actors can decrease the risk of a regional nuclear arms race.

I would like to congratulate George Perkovich, a vice president for studies at Carnegie, and Sinan Ülgen, a visiting scholar at Carnegie Europe, for putting together this collection of thoughtful and original essays on the nuclear policy of an important NATO ally at a critical cross-roads. The book provides a wealth of new information and thoughtful commentary to help policymakers and experts understand Turkey's energy and national security policies and develop strategies to bolster the nonpro-liferation regime.

William J. Burns
President
Carnegie Endowment for International Peace

WHY TURKEY?

SINAN ÜLGEN

In regions where nuclear weapons are deployed and pressing security dilemmas exist, states' nuclear policies often have, or could have, multiple dimensions. States in such circumstances may have civilian nuclear programs to provide energy or isotopes for medical and agricultural purposes. These civilian programs may be active and ambitious, or fledgling and focused on research and development. States in challenging security environments may also pursue military nuclear policies to deter potential adversaries. These policies may focus on alliance relationships whereby some states produce and control nuclear weapons, extending deterrence on behalf of their allies, which in turn may share responsibility by hosting nuclear weapons and participating in alliance policy planning. Or, states may seek to develop the option to produce their own nuclear weapons in the future, albeit with great complication due to the constraints that apply to parties to the Treaty on the Non-Proliferation of Nuclear Weapons (NPT).

The trajectories of states' nuclear policies sometimes are uncertain. A civilian nuclear program may always remain purely peaceful and thus not merge into a military nuclear program. Or, what began as a civilian program could alter its trajectory and become part of a policy to acquire a nuclear deterrent. Between these two trajectories, a state may seek a hedge by acquiring some dual-use capabilities that are not excluded for non-nuclear-weapon states under the NPT, without deciding to build nuclear weapons. Since 1970 when the NPT entered into force, only Libya, Iraq, Iran, North Korea, and possibly Syria have deviated from their obligations as parties to the treaty and sought to develop nuclear weapons.[1] The nuclear programs in Iraq, North Korea, and Iran began, at least ostensibly, as civilian efforts and were diverted to military purposes.

If a state decides to acquire its own nuclear weapons—as distinct from relying on an extended deterrent provided by an ally or allies—it will need a significant array of nuclear expertise and technological capabilities. Given that all non-nuclear-weapon states today are parties to the NPT, a state wanting to acquire nuclear weapons would probably seek to develop the necessary expertise and technical resources under the guise of a peaceful civilian nuclear program. To do otherwise—to abruptly undertake a crash nuclear-weapon program—would trigger alarms and countervailing action by the international community that would pose extreme risks to such a state. Alternatively, to seek to procure nuclear weapons from another state would pose other risks, and to be feasible would also require the prior development of expertise and infrastructure to make use of procured weapons and maintain them safely. Thus, an extensive civilian nuclear program, including the capacity to enrich uranium and/or extract plutonium from spent fuel, would appear necessary as a precursor for acquiring nuclear weapons. Such a program, in turn, would signal defenders of the nuclear nonproliferation regime to cast a wary eye.

Turkey exists in a nuclearized environment fraught with security dilemmas. Indeed, Turkey hosts several American nuclear bombs on its territory as part of the North Atlantic Treaty Organization (NATO) deterrent. Along Turkey's southern border, Iraq had a major clandestine nuclear weapon program, Syria had a secretly under-construction plutonium production reactor, and Iran has been detected violating its International Atomic Energy Agency (IAEA) safeguard obligations and conducting nuclear-related activities with possible military dimensions. Two of these neighbors, Syria and Iraq, have used chemical weapons in conflict, and Iran is believed to have the capacity to produce both chemical and biological weapons.

Turkey is a non-nuclear-weapon state in good standing under the NPT, with a robust civilian research and development program and extensive plans to host nuclear power plants on its territory. These plans call for established nuclear reactor vendors to build and operate plants in Turkey for many years before transferring them to Turkish operators who will gain the necessary expertise and experience in the intervening period. Turkey also wants to take advantage of its foray into nuclear energy to develop its own technological capacity. So Ankara has been interested in the transfer of nuclear technology and has not ruled out an interest in developing an indigenous capability to enrich uranium. All of this reasonably fits the profile of an advancing state and society with a robust modern economy whose need for diversely supplied electricity will grow significantly.

The trajectory of Turkey's nuclear energy policies and capabilities may never divert from a purely civilian course. But, noting the security environment in which Turkey lives, and the uncertain evolution of NATO and U.S. security guarantees, Turks and some international observers naturally speculate on the possibility that someday military-security considerations could cause the trajectory of Turkey's nuclear program to veer toward an altered line of Turkish security policy.

This book begins by describing the current status and trajectory of Turkey's civilian nuclear policy and program. In almost all countries, and in much of the international media, projections of the growth of nuclear energy prove vastly exaggerated. This is natural: proponents, including vendors, have incentives to understate the total costs of nuclear power plants and related infrastructure and the time it takes to construct them. Politicians and reporters not well-versed in the complicated realities of nuclear energy often do not know how to scrutinize such claims.

The first chapter by Gürkan Kumbaroğlu offers a welcome corrective by providing a historically informed, balanced, and incisive analysis of Turkey's burgeoning nuclear power program. Kumbaroğlu underscores the need for nuclear power by drawing attention to the high and continuing demand for electricity generated by a growing economy and a large manufacturing sector. He also analyzes the economics of the country's first nuclear power plant to be built by Rosatom in Akkuyu on Turkey's

Mediterranean coast. He concludes, after reviewing different risk categories, that the unique Build-Operate-Own investment model appears to be economically advantageous for Turkey with the predetermined average purchasing price of electricity from the nuclear power plant being significantly lower than today's average electricity wholesale price in Turkey.

In the second chapter, İzak Atiyas focuses on the state of play regarding the regulation of nuclear power in Turkey. He seeks to set out the desirable features of a regulatory framework to ensure the safety of nuclear energy in Turkey. Atiyas maintains that the quality of the regulatory framework is closely related to the degree of independence and to the extent to which mechanisms ensuring transparency and accountability are in place. He states that the current regulatory authority, and the regulatory framework in general, does not yet satisfy the requirements of independence, transparency, and accountability. Looking ahead, he asserts that Turkey is faced with two important and highly interrelated tasks. One is that Turkey needs to complete the legislative infrastructure for nuclear energy. The other task is highly political in nature as it requires the political authority to delegate substantial power to an independent authority.

The third chapter by Doruk Ergun can be seen as an introduction to the strategic dimension of the debate on nuclear policy. In this chapter, Ergun reviews Turkey's security strategy since the end of World War II and provides a context for the discussion of the role of nuclear weapons in Turkey's security strategy. He examines the evolution of NATO's nuclear deterrence and the role of forward-deployed tactical nuclear weapons in Europe. He then analyzes the post–Cold War security environment and proliferation threats surrounding Turkey and sets out Ankara's response to the perceived regional threat of weapons of mass destruction (WMD). Ergun concludes by stating that although Turkey continues to be faced with other asymmetric and conventional threats both within and beyond its borders, Ankara will not risk the diplomatic, political, economic, and military fallout from seeking nuclear weapons unless it is convinced that NATO security guarantees have completely eroded and that Turkey is compelled to fend for itself.

In the fourth chapter, Can Kasapoğlu provides the backdrop for Turkey's strategic thinking on the role of tactical nuclear weapons. He remarks that Turkey's nuclear stance within NATO has generally been seen through the prism of national prestige and the consolidation of North Atlantic security ties. He maintains, however, that deeper insight into the Turkish case and Ankara's delicate military strategic balance in its hinterland with respect to strategic weapon systems uncovers a more complex rationale beyond the generally accepted motive of commitment to NATO's burden sharing and nuclear posture. Kasapoğlu argues that the positioning of NATO tactical nuclear capabilities on Turkish soil has more than a symbolic meaning for Ankara and that Turkey's nuclear stance within NATO is linked to threat perceptions emanating from strategic weapons proliferation in the Middle East.

In the fifth chapter, Aaron Stein reviews Turkey's efforts to improve its defensive and offensive capabilities against the threat of WMD. Stein argues that Ankara has expressed a sustained interest in procuring offensive and defensive missile systems that are intended to work together to augment Turkey's capabilities to target asymmetric threats and to bolster the country's defense against a ballistic missile attack. He indicates that Turkey is pursuing ballistic and cruise missiles and is eager to complement these capabilities with a robust intelligence-gathering capability that relies on space-based and unmanned systems that are intended to work together to provide a better defense against regional missile proliferation. He asserts, however, that these plans are not tied to Turkey's civilian nuclear efforts and do not appear aimed at providing Turkey with a nuclear-capable delivery system. For Stein, the objective of the missile program appears to be to deepen the country's ability to target ballistic missiles before they are fired and to provide Turkish military planners with greater long-range conventional strike capabilities against a variety of targets.

In the sixth chapter, Mark Hibbs describes the international institutional framework in which Turkey's civilian nuclear program operates. This framework has been established to foster the peaceful development of nuclear energy while preventing nuclear-weapon proliferation and, ultimately, creating conditions for nuclear disarmament. To the extent that motivations for proliferation and incentives for disarmament depend on how states' leaders perceive their political-security environment and the costs and benefits of alternate courses of action, this chapter provides an apt bridge to the next chapters, which explore Turkey's perceptions and policies regarding nuclear deterrence and national power and security.

In the seventh chapter, Mustafa Kibaroğlu explores how Turkey would seek to develop its own nuclear weapons if it wanted to. In the first part of the chapter, Kibaroğlu discusses factors that to date have kept Turkey from seeking to produce its own nuclear-weapon capabilities. He reviews Turkey's domestic interests and characteristics as well as a number of international factors that have constrained Turkey's possible ambitions, such as its membership in NATO, its adherence to the nuclear nonproliferation regime, and its European Union (EU) vocation. In the second part of the chapter, Kibaroğlu speculates about possible courses of action that Turkish policymakers could adopt in case they decided to acquire nuclear weapons. He discusses a number of critical issues such as who should be responsible for the development of a nuclear-weapon program, what should be the strategy for evading the country's commitments under the nuclear nonproliferation regime, which capabilities and technologies should be acquired and/or indigenously developed for becoming self-sufficient in the long term, and who should be Turkey's international partners in this effort. Kibaroğlu concludes that there are no feasible scenarios under which Turkey could expect to effectively use its nuclear power status, if and when it is achieved. However, there are scenarios in which Turkey's vital interests would be seriously damaged simply because the country would have attempted to acquire a nuclear-weapon capability.

In a follow-up chapter, Jessica Varnum examines the trends and trigger events that could alter Ankara's nuclear proliferation decisionmaking. Varnum discusses the effects of domestic politics, the security environment, multilateral interests, and commitments, as well as current and projected capabilities on Turkey's nuclear future. Chapter 8 should thus be read together with the previous chapter insofar as both seek to understand how Turkey could be diverted from its current path of having a clean track record in the area of nonproliferation. Varnum states that many such dynamics could influence the direction of Turkish nuclear policymaking. However, she maintains that only rarely could a single "trend or trigger event" cause a country otherwise predisposed to nonproliferation to actually launch a nuclear-weapon program. Varnum concludes by asserting that in Turkey's case, a range of factors, including domestic politics, the security environment, multilateral interests and commitments, and current and projected nuclear capabilities, favors continued proliferation restraint.

Finally, in the conclusion, George Perkovich highlights this volume's purpose of creating a baseline of current information and analysis on Turkey's history, interests, capabilities, and dilemmas that we hope analysts, journalists, and policymakers will build upon. To this end, Perkovich emphasizes several clusters of facts and insights from these chapters that may correct assumptions or assertions on the part of some casual observers and commentators regarding Turkey's efforts to pursue nuclear energy and national security. He concludes with a set of questions whose answers may predict the trajectory of Turkey's nuclear future.

NOTE

1   Iran argues that it has never sought nuclear weapons. The United States and others contend that before 2003 there was a nuclear-weapon program in the country. The IAEA has not rendered a judgment, but it has identified a number of activities that have possible military dimensions, making it impossible, without further information, for the agency to determine that Iran's nuclear program has been and remains exclusively peaceful. Syria was secretly developing a reactor that the IAEA and other states believe was meant to produce plutonium, presumably for subsequent use in a nuclear-weapon program.

CHAPTER 1

TURKEY AND NUCLEAR ENERGY

GÜRKAN KUMBAROĞLU

TURKEY'S LONG-STANDING INTEREST IN NUCLEAR POWER

Turkey's quest for nuclear energy can be traced back almost as far as its membership in the Atoms for Peace program in 1955. It was in the following year that Turkey's Atomic Energy Commission was founded under the auspices of the prime ministry. A 1 megawatt (MW) research reactor began operation in 1962. The Ministry of Energy and Natural Resources was established a year later, and the Electric Works Study Administration, a division within the ministry, began feasibility studies for the construction of a nuclear power plant: a 300–400 MWe (megawatts of electrical output) pressurized heavy water reactor (PHWR) that was planned to go online in 1977. The studies lasted until 1970, but the project was later canceled due to issues related to site selection, among others. Meanwhile, in 1972 a Nuclear Power Plants Division was set up under the auspices of the Turkish Electricity Authority (TEK), which conducted feasibility, bid specification, and site selection studies for a 600 MW power plant that was planned to come online in 1983. With respect to site selection, three places ranked highest on the cost-benefit scale: Akkuyu in Mersin Province, Inceburun in Sinop Province, and İğneada in Kirklareli Province.

Akkuyu was picked as the site of Turkey's first nuclear power plant and received its site license in 1976. A Swedish consortium consisting of ASEA-ATOM (charged with the boiling water nuclear reactor) and STAL-LAVAL (charged with the turbine) was chosen in 1977 for construction of the 600 MW plant. However, the negotiations were stopped in 1979 after the Swedish government refused to provide financial guarantees and the Turkish government failed to display strong political will on the issue due to political instabilities at the time that resulted in the 1980 military coup.

A renewed effort for nuclear energy took place in the 1980s. Site selection studies for a second nuclear power plant began in 1980, and Inceburun was selected by the Nuclear Power Plants Division. In 1983, Ankara sent letters of intent to three companies for the construction of three or four nuclear power plants in Turkey. One letter went to the West German Siemens-Kraftwerk Union for the construction of a 990 MWe pressurized water reactor (PWR) in Akkuyu, another to Atomic Energy of Canada Limited (AECL) for a 655 MWe CANDU reactor, and the last one to General Electric (GE) in the United States for the construction of one or two boiling water reactors in Inceburun with a total installed capacity of 1,185 MWe.[1] However, after preliminary site studies in Inceburun revealed that the likelihood of earthquakes in the area increased the cost of constructing a facility,[2] GE determined that it would not be feasible to build a nuclear power plant there until further comprehensive seismic studies were made and halted the negotiations. Negotiations with the other companies that began in 1984 were not fruitful either. Ankara's preferred financing model, Build-Operate-Transfer (BOT), was one reason for this. In the BOT model, the contractor company pays for the construction and operating costs of a given facility and operates the facility for a predetermined period of time (fifteen years in this case), thus recouping its expenses plus profit, after which it transfers control of the facility to the host government. At the time, the BOT model had never been used in a nuclear power plant. Initially, Siemens-Kraftwerk Union pulled out of the negotiations due to disagreements on the financing and partnership arrangements. Negotiations with AECL continued, and the Turkish parliament ratified a nuclear cooperation agreement with Canada in 1985. AECL and the Turkish authorities reportedly signed a preliminary agreement, in which it was agreed that TEK would have 40 percent of the shares and that AECL and its partners would have the remaining 60 percent.[3] But the talks broke down over issues regarding financial guarantees, as neither the Turkish government nor the Canadian government wished to provide extensive financial guarantees. Moreover, AECL requested that "risk coverage" should be specified in the contract and that during the time it operates the facility, Turkish electricity purchases should be in dollars and should be sufficient for AECL to recoup its expenses.[4] By 1987, the talks with the Canadian side had collapsed.

The 1980s were characterized by capacity building at the institutional level as well as by the ratification of various international agreements. In 1980, Ankara ratified the Treaty on the Non-Proliferation of Nuclear Weapons (NPT), which it had signed in 1969. In an agreement signed with the International Atomic Energy Agency a year later, Ankara accepted IAEA safeguards and supervision over all existing and future nuclear facilities. A year after that, in 1982, Turkey's Atomic Energy Commission was reorganized as the Turkish Atomic Energy Authority (TAEK) and placed once again under the auspices of the prime ministry. In 1983, another organization under the name Nuclear Power Plants Institution was founded and put in charge of managing various aspects of nuclear power generation, such as constructing and managing plants, building necessary infrastructure, conducting feasibility studies, and so on. It would later turn out that the organization existed only on paper and was shut down in 1991. In 1984, Turkey became a member of the Nuclear Energy Agency of the Organization for Economic Cooperation and Development (OECD).

The Chernobyl disaster in 1986 compelled the Turkish government to suspend its nuclear energy ambitions. In 1988, the Nuclear Power Plants Division was disbanded during the reorganization of TEK and most of the staff left TEK, taking their years of expertise with them.[5] In the same year, Turkey and Argentina signed a fifteen-year nuclear cooperation agreement, which was ratified by the Turkish parliament in 1992.[6] Ankara was interested in two nuclear reactors of Argentinean design: the 380 MWe Argos PWR and the 25 MWe CAREM-25. The two countries agreed to establish a joint architecture-engineering firm in 1990 and committed to building two CAREM-25 units, one in each country, in a deal in which Turkey would take the lead in financing the plants and Argentina would take the lead in providing the technology.[7] It was expected that if the cooperation in CAREM-25 bore fruit, the 380 MWe Argos PWR might follow it. However, the Argentinean project was also canceled, reportedly because the United States, the Soviet Union, and other countries had proliferation concerns tied to the CAREM-25 deal. Executives at TAEK concluded that going ahead with the Argentinean project might hamper Turkey's chances of cooperating with other countries on nuclear issues in the future.[8]

In 1992 the Ministry of Energy and Natural Resources submitted a report to the government in which it stated that unless new energy sources were installed before 2010, the country would face an energy crisis. The report strongly suggested that nuclear energy should be taken into consideration.[9] Turkey's annual energy consumption growth rate since 1975 had been around 8 percent,[10] a trend that has continued since then. In the same year, TEK sent letters to prominent nuclear companies asking for technical and financial information on a 1,000 MW nuclear power plant consisting of one or two units that would come online in 2002 and be built with the BOT model. The following year, nuclear plants were included once again in Ankara's investment program, and electricity generation through nuclear power plants was listed as the third-highest priority of the country by the Science and Technology Upper Council of the Scientific and Technological Research Council of Turkey (TÜBİTAK).[11] Ankara started taking offers from consulting companies in 1994. A major structural change came into effect the same year, namely the split of TEK into the Turkish Electricity Distribution Co. and Turkish Electricity Generation Transmission Co. (TEAŞ), which retained authority over nuclear matters and later reestablished a Nuclear Power Plants Division.

Revised tender specifications for the plant in Akkuyu were released at the end of 1996, and bids from three companies were taken the following year: for two 669.5 MW or four 665.5 MW CANDU type PHWRs from AECL; for one or two 1,482 MW PWRs from Nuclear Power International, which consisted of Siemens and Framatome, a French firm; and for a 1,218 MW PWR from the consortium of Westinghouse from the United States and Mitsubishi from Japan.[12] The government, however, delayed its decision "no less than eight times"[13] between 1998 and

2000 and finally abandoned the plans in July 2000 due to economic cir-cumstances. Furthermore, the Nuclear Power Plants Division was shut down once more.

Turkey signed several nuclear cooperation agreements in the second half of the 1990s:[14] with Germany in 1998 (not ratified), with South Korea in 1998 (ratified in 1999), with France in 1999 (ratified in 2011), and with the United States in 2000 (ratified in 2006). In 2008, after nearly six decades of wavering, it appeared that Ankara had become more resolute with regard to its nuclear energy program. The site in Akkuyu was once again opened to the bidding process, for which only one bid was submit-ted in 2008 by a consortium of 14 parties including two Russian compa-nies, Atomstroyexport and Inter RAO, and a Turkish firm, Park Teknik, for four VVER-1200 (each 1,200 MWe) reactors. Even though high-level talks were conducted the following year (including some with the Russian prime minister, Vladimir Putin) and two nuclear cooperation agreements were signed in August 2009, the deal collapsed. One reason for the failure to consummate the deal was the high prices for proposed electricity sales.

The sides then chose to skip the bidding process in favor of direct talks between the respective governments. On May 12, 2010, an agreement on cooperation between the government of the Republic of Turkey and the government of the Russian Federation was signed in relation to the con-struction and operation of a nuclear power plant at the Akkuyu site. It was ratified by the Turkish parliament on July 15 and the Russian parliament in November the same year. Under the agreement, Rosatom agreed to build four VVER-1200 reactors and to own and operate them, making Akkuyu potentially the first foreign nuclear power plant to be constructed under the Build-Operate-Own (BOO) investment model. TETAŞ, the Turkish Electricity Trading and Contracting Company, was assigned to purchase 70 percent of the electricity generated by the first two reactors and 30 percent of the electricity generated by the next two[15] at a weighted average price of 12.35 cents per kilowatt-hour (kWh), not including value-added tax (VAT), for fifteen years starting from the date of commercial operation for each power unit. After the first fifteen years, the project company is to sell the electricity on the open market and transfer 20 per-cent of its profits to the Turkish government.

Although Russia has agreed to fully finance the project, the estimated costs are on the rise. In 2012, the project company suggested that the plant would cost $18.7 billion; Rosatom later said that this figure could rise as high as $25 billion.[16] In fact, the Russian minister of energy, Alexander Novak, said in October 2013 that the government will be selling some shares of Inter RAO, one of the country's largest public energy companies, to finance the project.[17]

Meanwhile, several important developments also took place with regard to a second nuclear power plant at the Sinop site. In 2010, the Turkish Electricity Generation Corp. (EÜAŞ) signed an agreement with the Korea Electric Power Corp. (KEPCO) to prepare a bid for the construction of four APR-1400 reactors scheduled to come online in 2019. Yet this proposal proved to be short-lived, reportedly because the South Korean company insisted that Ankara should provide treasury loans and guarantee electricity sales proceeds directly from the government instead of through TETAŞ. Also, the sides could not agree on the power prices.

Ankara then turned to Japanese companies for the deal and signed an agreement at the end of the year for the preparation of a bid. Toshiba and Tokyo Electric Power Company were involved in talks to construct four 1,350 MWe advanced water boiling reactor units,[18] but the talks were suspended at the request of the Japanese side due to the Fukushima incident in 2011. Talks continued throughout 2011–2013, during which Turkey considered offers from Canadian, South Korean, Chinese, and Japanese bidders. Turkey eliminated Candu Energy Inc. from the process in 2012, KEPCO in early 2013, and finally China in April 2013.

On May 3, 2013, Ankara chose a consortium led by Japan's Mitsubishi Heavy Industries and France's GDF Suez. They will be building four Atmea1 reactors, which will most likely be the first of their kind[19] and amount in total to 4,400 MW. Construction is planned to start in 2017, with the reactors expected to come online in 2023, 2024, 2027, and 2028.[20] The electricity purchase agreement is to be effective for a twenty-year period after each reactor comes online. Turkey would then buy electricity at a fixed price of 11.83 cents/kWh (nuclear fuel included) or 10.80 cents/kWh (fuel not included), giving it the option to obtain the fuel from other sources.

In the Sinop project, Ankara revised its financing policy, which had been a major obstacle in its numerous attempts at nuclear power, and decided to acquire shares of the project. Loans from the Japan Bank for International Cooperation will cover 70 percent of the expected $22 billion cost of the project, while the remaining 30 percent will be covered by equity. EÜAŞ will provide 25 percent of the equity and will be a partner to the project at the same percentage. Reportedly, EÜAŞ will also have 49 percent of the shares of the project company but may opt to sell 24 percent of its shares in time. The main operator will be GDF Suez, a French multinational electric utility company.[21]

At the time of writing, not much was clear about Turkey's ambitions for a third nuclear power plant. According to a news report from 2012, some of the locations that the Turkish government is considering are Ankara-Nallıhan, Kırıkkale-Nevşehir, and Beyşehir-Seydişehir in Central Anatolia, and Akçakoca-Ereğli and Kırklareli-İğneada close to Istanbul and the industrial centers in the Marmara region.[22] Taner Yıldız, the minister of energy and natural resources, suggested in October 2013 that the government still had not decided on the location and that Ankara has asked Japan to conduct studies on four or five sites.[23]

Obviously, in terms of nuclear energy, Turkey has managed to make much more headway since 2010 than in the preceding five decades. One reason for this major change in decisiveness has been the economic stability and growth that the country has enjoyed for the past decade. The economic restructuring after the financial crisis in 2001 has allowed Turkey to enjoy a GDP growth rate averaging more than 5 percent between 2002 and 2011—which is reflected in the high rates of increasing energy demand. Another change in the past decade is due to the landslide victories of the Justice and Development Party (AKP) in the last three general elections. This relative political stability has allowed subsequent governments to put more political will behind their nuclear policies and follow them without much internal challenge and debate.

During this period of renewed activism, Turkey also has increased its cooperation with outside actors. A civil nuclear cooperation agreement with the United States entered into force in 2008. The country also signed nuclear cooperation agreements with South Korea in June 2010, with Japan in December 2010, with Jordan in February 2011, and two more with China in April 2012. Moreover, Ankara has decided to take part in the EU stress test program for nuclear power plants[24] and signed an agreement with the IAEA in November 2012 for cooperation in the development of its nuclear program infrastructure. The IAEA conducted an Integrated Nuclear Infrastructure Review in November 2013 at the request of Turkey[25] and issued a largely positive report in February 2014.

WHY DOES TURKEY SEEK NUCLEAR ENERGY?

As is evident from this historical narrative, there is significant political will in Turkey to phase in nuclear power. The AKP government has put nuclear energy generation on its ten-year agenda, the much advertised "2023 vision," which contains a set of ambitious goals to be completed by the centennial of the Republic of Turkey. Yıldız, the energy minister, has declared that the country plans to have two power plants in operation and a third one under construction by 2023.[26] The aim is to reach a nuclear power installed capacity level of at least 10,000 MW by 2030.[27] One of the key drivers of Turkey's ambitious nuclear motivation is the strong increase in demand.

Electricity supply considerations in Turkey have been spurred by rapid growth on the demand side and the historical dominance of hydropower and fossil-fuel–based thermal power generation on the supply side. Demand has been growing at a remarkable rate, averaging 8 percent per year over the past decade. It is expected that the increase in electricity demand will continue as the economy grows, albeit at a slightly lower rate. Official projections assume an annual growth rate of nearly 6.5 percent in the low-growth scenario over the next decade. In other words, a doubling of electricity demand is expected within the next ten years. In terms of supply capacity, the rapid growth of demand translates to investment requirements of at least 25,000 MW over the next decade. The most recent official projections, published in June 2013, predict that gross peak summer demand will grow from 38,159 MW in 2012 to 74,429 MW in 2022, while the winter peak will increase from 36,812 MW in 2012 to 64,918 MW in 2022.[28] Assuming an average growth rate of 7.5 percent per year for the first five years and 6.4 percent for a ten-year period (both starting in 2013), per capita electricity demand is forecast to grow from 3,224 kWh in 2013 to 5,400 kWh in 2022. Even these figures could be considered conservative, as they are far below the OECD average of 8,226 kWh in 2011.[29]

Meeting the growing demand to ensure a secure supply of electricity at affordable prices and in an environmentally sustainable way appears to be a challenge for Turkish policymakers, especially because of limited indigenous capacity of cheap and clean energy sources. Utilizing domestic resources is a policy priority as Turkey's import dependency has been increasing rapidly over time, reaching 80 percent and feeding concern over supply security and price stability. Almost half of electricity generation (44 percent in 2012) depends on natural gas, which exacerbates import dependency given that 98 percent of natural gas is imported. Turkey's import of mineral fuels, lubricants, and related materials in 2012 was $60 billion, making up 25 percent of total imports. The country's short-term foreign debt stock was $101 billion in 2012, and its medium- and long-term foreign debt stock was $226 billion. Foreign currency inflow is needed to pay back that debt. However, Turkey's imports surpass exports, causing a current account deficit and preventing net foreign currency inflow. The energy import bill of roughly $60 billion exceeds the 2012 current account deficit of $47 billion. In other words, if it were possible to reduce energy imports by 80 percent, Turkey's balance of payments would yield a surplus. Evidently, the $60 billion worth of energy imports constitutes a major burden on the Turkish economy.

There are, however, various obstacles that limit a wider utilization of indigenous reserves. In terms of fossil fuels, only low-quality coal is available, and its use faces criticism because of pollutant emissions. Furthermore, construction of coal-fired power plants without carbon capture and storage conflicts with the national climate change strategy, which argues for the expansion of low and zero greenhouse gas emission technologies. Nuclear power appears to be an attractive option both in terms of import dependency and greenhouse gas mitigation. Nuclear energy generation, it is true, will create some level of foreign dependency due to the need to import the technology itself in the acquisition phase of the investment, as well as the need for uranium enrichment as a fuel in the operational phase. However, the possibility of know-how transfer, plus the fact that fuel accounts for only a small share of total generation cost (and the ability to acquire the long-term fuel requirements early on), weakens concerns over import dependency related to nuclear power generation. The prospect of know-how transfer and accumulation of knowledge, experience, and expertise appears to be a motivation for policymakers to embrace nuclear power. In terms of greenhouse gas mitigation, nuclear power is particularly attractive as the electricity generation it provides is carbon-free.

Climate change mitigation is a major challenge for Turkish policymakers, as the country's greenhouse gas emissions have been skyrocketing. With a growth rate of 115 percent during 1990–2010, Turkey had the fastest-growing emissions among Annex 1 parties to the United Nations Framework Convention on Climate Change. Malta was a distant second at 49 percent, and Australia was third at 30 percent. Carbon dioxide emissions resulting from power generation in Turkey grew fastest among all sectors—a remarkable 252 percent since 1990. Naturally, the growth of carbon emissions resulting from electricity generation is a result of technological choices. Had the investments for additional capacity been made purely on zero carbon technologies such as renewables and nuclear, there would have been zero growth of emissions from electricity generation.[30]

While the share of hydroelectric energy in total electricity generation decreased from 40 percent in 1990 to 25 percent in 2010, the share of thermal plants rose from 60 to 74 percent and wind power obtained a 1 percent share. Although the share of hydropower declined, its installed capacity level actually increased (by 2.3 times in twenty years) but was overshadowed by the faster increase of thermal power plants (by 3.4 times over the same period), leading to more carbon-intensive power generation. A technological shift toward less carbon-intensive power generation is considered key to reduce Turkey's rapidly increasing greenhouse gas emissions. The country's national climate change strategy document for 2010–2020 places nuclear energy as a zero emission technology among the mid-term activities to reduce greenhouse gas emissions. Additionally, the prodding of nuclear energy is included in Turkey's Climate Change National Action Plan for 2011–2023: nuclear power is being considered as an effective option to reduce the country's remarkable growth of greenhouse gas emissions.

There is some public opposition to nuclear power in Turkey by those who favor renewable power to meet increasing demand. Indeed, in terms of wind and solar power potential, Turkey appears to have an advantage over most European countries. The Ministry of Energy and Natural Resources presents the potential of wind power capacity in regions with wind speed of at least 7.5 meters per second as 48,000 MW (the current installed capacity of wind power is nearly 2,500 MW). The potential for solar power generation (whose current installed capacity is negligibly small) is also huge, as Turkey's least promising region appears to have a higher potential (in terms of both intensity and duration) than many European countries that have installed a considerable amount of solar power capacity. Hydropower remains partly unexploited as well. According to the State Hydraulic Works, Turkey has a potential of an additional 20,000 MW capacity of hydropower. With all this potential, the obvious question arises: why should Turkey turn to nuclear power when such a great unexploited renewable power generation potential exists in the country? The response by policymakers emphasizes two points: it is primarily due to economic reasons and secondarily due to technical restrictions (grid constraints) that renewables are not considered as a solution to meet all of the additional demand over the next decade. Indeed, both arguments have solid footing, as the intergovernmental agreement with Russia for Turkey's first nuclear power plant at Akkuyu exhibits a rather extraordinarily low price and the capacity of the Turkish grid is limited.

The Need for Low-Cost Power Generation

It is particularly important for Turkey to introduce cheaper alternatives into the power generation mix, as end-use electricity prices have been far above the OECD average for the past decade. Figure 1 shows the historical development of Turkish end-use electricity prices for households and industry compared with the OECD total, based on data from the International Energy Agency and converted using purchasing power parities. The gap indicates the relatively high level of end-use prices in Turkey and makes explicit the importance of introducing cheap generation. Under these circumstances, cost reduction emerges as a priority goal along with security of supply. The agreement with Russia appears to be a win-win in the sense that it not only contributes to supply security but also offers a guaranteed sale at a price that is significantly lower than today's average wholesale price in Turkey. Hence, the country's first nuclear power plant features low-cost power supply made in Turkey using Russian technology. It will help to reduce the price gap shown in figure 1. In the case of nuclear power, safety considerations are of paramount importance and should not be weakened to reduce cost. Subsequently, the agreement is critically elaborated from various aspects ranging from the investment model to safety and security issues.

FIGURE 1. End-Use Electricity Prices (using purchasing power parities)

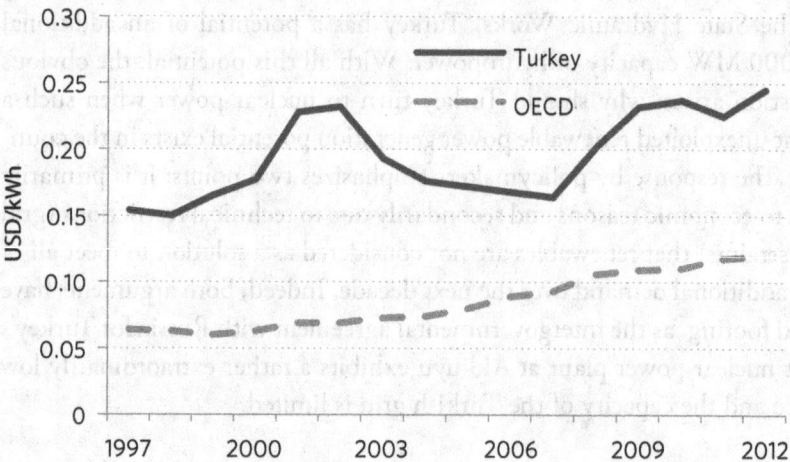

Source: International Energy Agency, "Energy Prices and Taxes Quarterly Statistics" (Second Quarter 2013).

Limitations of Grid Capacity

Using the grid for balancing renewable power generation will require transport of electricity from areas with high renewable potential to the demand centers. Such a grid will require large amounts of excess transmission as well as reserve capacity to avoid congestion and blackouts as the intermittent electricity surges shift from one minute and one area to the next. Investments in the quality and capacity of the grid are needed to keep pace with investment in intermittent renewable generating capacity. For wind power, the most promising and competitive renewable energy source in Turkey, a capacity level of 12,369 MW is considered feasible by the Energy Market Regulatory Authority. This would correspond to nearly 33 terawatt-hours (TWh) of power generation, assuming an average capacity factor of 30 percent. The 33 TWh of additional generation is far from meeting the incremental demand, which is forecast to be at least 150 TWh until 2020 (according to official estimates in a low-growth scenario). For solar power, which is the runner-up among renewables in Turkey, grid capacity to accommodate solar power has been declared to be much less: an upper limit of 600 MW (which would generate less than 1.5 TWh) has been announced for solar power generation that can be connected to the grid, which has received applications totaling 7,094 MW as of 2014. The amount of investment into the grid that would be needed to accommodate more intermittent power is not known. Obviously, under the current grid conditions, the addition of non-intermittent power generation is inevitable to meet growing demand. Given that other thermal power generation technologies are associated with carbon emissions or import requirements, nuclear power emerges as a desirable option for security of supply under grid conditions where a sufficient amount of renewables cannot be accommodated.

In addition to all these aspects, the Turkish preference to import nuclear technology seems to be motivated by the fact that most developed economies set an example by making extensive use of nuclear power generation and make it appear as if nuclear technology has been a crucial element in their economic development. That said, changes in policy after the Fukushima incident, such as the decision by Germany to phase out nuclear power, have not yet proved feasible.

TURKEY'S FIRST NUCLEAR POWER PLANT: WHY RUSSIA?

An Economic Evaluation

The Turkish Electricity Trading and Contracting Company has guaranteed the purchase of 50 percent of the power generated from the Russian plant to be built in Akkuyu over a fifteen-year power purchase agreement at an average price of 12.35 cents per kWh, excluding VAT, in nominal terms. To compare this nominal figure with current prices, a discount rate has to be applied. Accordingly, the real value in today's purchasing power (that is, discounted back fourteen years, from 2027 to 2013) is computed under different discount rate assumptions in the range of 3.25–7.13 cents per kWh as shown in table 1.

TABLE 1. Akkuyu Nuclear Power Plant Agreement Average Purchase Price (2020–2035) in Real and Nominal Terms

PRICE [U.S.¢/kWh]			
2013 real			**2027 nominal**
LOW DISCOUNT RATE (4% per year)	MEDIUM DISCOUNT RATE (7% per year)	HIGH DISCOUNT RATE (10% per year)	
7.13	4.79	3.25	12.35

Source: Author's calculations.

On December 18, 2012, the average wholesale price for electricity (to be used in determining the renewable power subsidies in 2013) was declared by the Turkish Energy Market Regulatory Authority as 8.67 cents per kWh. Obviously, the agreed-upon price level for nuclear power generation at the Akkuyu plant is significantly lower than today's average wholesale price in Turkey. Technology-specific comparison of the agreed-upon purchase price with other reactor-level international data indicates that the Russians have set a very low profit margin, if any.[31] Furthermore, after the expiration of the power purchase agreement (not earlier than fifteen years after the start of commercial operation), Russia shall give 20 percent of the net profit on a yearly basis and throughout the lifetime of the project to the Turkish party. In summary, given that the Turkish side bears no financial risk, the Russian agreement appears to be exceptionally economically advantageous for Turkey.

The Investment Model

For the Akkuyu project, an intergovernmental agreement was signed in May 2010 between Turkey and Russia and ratified by the parliaments of both countries. As noted earlier, the investment model based on a Build-Operate-Own scheme puts almost all financial risk on the Russian side: TETAŞ guarantees to purchase 50 percent of the total production for fifteen years at an average weighted price of 12.35 cents per kWh, excluding VAT. The annual differences will make it possible for the project to be reimbursed and shall be calculated in a manner so as to not surpass the upper limit of 15.33 cents per kWh. It is assumed that this price will pay back the investment expenditures of the project company in fifteen years. In case the project revenue fails to pay back the cost, it is the project company's responsibility to cover additional financing needs. As such, the investment model appears to be attractive for Turkey in that there is no risk of additional financial burden in case of a cost overrun, assuming that the guaranteed purchase agreement becomes valid. However, other problems might arise if the project company does not sign the agreement as discussed in the financial risk section below. Liabilities as well as waste management, decommissioning, and other major issues also are the responsibility of the project company, as discussed below.

Russia's Nuclear-Waste Advantage

According to the agreement, the project company shall make the necessary payment for waste management to a relevant fund as stipulated by the applicable Turkish laws and regulations. Accordingly, it will contribute 0.15 cents per kWh to a fund established for financing nuclear-waste management. Furthermore, according to the agreement, the project company is responsible for performing the waste management activity. Accordingly, it can be expected that the fund's accumulated payments will be paid back to the project company to cover the cost of waste management. It is mentioned in the agreement that spent fuel of Russian origin may be reprocessed in the Russian Federation subject to a separate agreement that may be reached by the parties. For Turkey, the Russian agreement appears to be advantageous in terms of waste management because it is the responsibility of the project company and the spent fuel will be transported back to Russia for reprocessing or final disposal. However, the means and safety and security aspects of transporting nuclear fuel are yet unclear and need to be determined.

The Financial Risk

As mentioned, all financial risk related to a possible cost overrun is on the project company since the agreed-upon weighted average sales price of

12.35 cents per kWh is fixed, not being affected by cost. Commercial motivation should assume that this price will be higher than the wholesale prices that prevail in the liberalized electricity market at that time (otherwise it will be subsidized by the project company). However, it should be noted that the parties may terminate the intergovernmental agreement at any time by way of one year's notice. The possibility of Russia's easily canceling the agreement can be considered a potential risk for the Turkish side as it may lead to a capacity gap with ensuing blackouts and price increases. This could occur if the wholesale prices were such that the project company would be better off canceling the agreement and selling the power generation in the market under competitive conditions. It would certainly be the case if the wholesale prices that prevail in the liberalized market turn out to be significantly higher than 12.35 cents per kWh. It could even be the case if market prices are less than 12.35 cents per kWh due to the market power potential of the project company.

Cost Minimization

All parties involved have motives to minimize cost. On the Turkish side, the fixed-price contract appears to be advantageous as it is low in real terms and carries no financial risk. On the Russian side, the fixed-price contract encourages the project company to reduce costs to the lowest level possible as this is the only way to increase profits. In the construction phase, this will lead to efforts to avoid delays and cost overruns. Although this is a positive aspect from an economic standpoint, it implies a potential danger specific to nuclear power generation. Many cost items in nuclear power generation projects are related to safety and security measures, and so scrimping on these measures can lead to significant cost savings—and risk. Therefore, recognizing the economic incentive of the project company to place a low importance on safety and security issues, strict control by an independent and competent authority to enforce all safety and security measures is essential. The regulatory and enforcement environment in Turkey, however, displays weaknesses, as outlined in an earlier study.[32]

Third-Party Liability

According to the intergovernmental agreement, third-party liability for nuclear damage will be regulated in compliance with the international agreements and instruments to which the Republic of Turkey is and will be a party, as well as national laws and regulations in Turkey. Currently, there is no upper limit on liability according to the Turkish law on obligations, and thus the responsibility of the project company is not limited in the event of an accident. However, an agreement to set an upper bound on liability may be forthcoming as it is an item under negotiation between the project company and the Ministry of Energy and Natural Resources. If Turkey ratifies the Amending Protocol to the Paris Convention, operator liability will have to be regulated to cover at least €700 million ($885 million).

In case the project company does not have the financial means to cover third-party liability and is not covered by insurance, the Russian government would be expected to assist as a last resort since the project company is a public one owned by the state.

Transfer of Know-How

According to the intergovernmental agreement, Turkish citizens are to be trained free of charge and widely employed for the purpose of operating the nuclear power plant. In fact, this transfer of know-how has aleady started: in order to implement the agreement, Russia provided an opportunity for Turkish students to receive specialized education in Russia. More than 9,000 students applied to the admission tests for studying at the Russian Research Nuclear University MEPhI (Moscow Engineering Physics Institute). Forty-eight students were selected in 2011, 69 students in 2012, and 100 students in 2013. They are trained in the city of Obninsk, Russia. The education program is planned to cover a total of 600 students.

Moreover, it is agreed that Turkish companies shall be widely employed for the supply of commodities, the rendering of services, and the implementation of works in connection with the construction phase of the project. However, it is also mentioned that the project company shall take into account the nature of the required work and any special safety requirements. As such, Turkish companies are not likely to be involved in the construction phase and supply of material for the core reactor. This is to be expected, as the know-how does not currently exist in Turkey, and it will not be economical for Turkish companies to produce reactor-specific parts to meet safety standards. Accordingly, Turkish companies can be expected to contribute to standard construction works, which will not lead to a transfer of know-how. Instead, the transfer of know-how is expected to occur predominantly through the training of Turkish students.

Decommissioning

According to the agreement, the project company shall make the necessary payment for decommissioning to a relevant fund as stipulated by the applicable Turkish laws and regulations. Accordingly, it will contribute 0.15 cents per kWh to a fund established for financing the decommissioning of the plant. Under the agreement, the project company is responsible for decommissioning the power station. Hence, the fund's accumulated payments would be paid back to the project company at the end of its lifetime to cover the cost of decommisioning.

TURKEY'S SECOND NUCLEAR POWER PLANT: WHY JAPAN?

On May 3, 2013, the Turkish prime minister and his Japanese counterpart signed an intergovernmental agreement on Turkey's second nuclear power plant project to be constructed in the Black Sea province of Sinop. Similar to the Russian plant in Akkuyu, the facility in Sinop is to comprise four reactors with a total installed capacity level of 4,400 MW (as announced by Mitsubishi Heavy Industries). Construction is scheduled to begin in 2017, with the first reactor to commence operations by 2023. According to the deal, a Japanese-French alliance led by Mitsubishi Heavy Industries will build the plant. GDF Suez will be the operator. As declared by the Turkish minister of energy and natural resources, Turkey will also have a stake in the nuclear power plant. It is expected that EÜAŞ will participate in the project with a share on the order of 30 percent. Negotiations on the conditions of the project are in progress, so no details are known at this time. It is therefore not possible to elaborate on the economics, investment model, risk, and other issues as has been done for the Akkuyu plant. One immediate advantage can be mentioned, however. Because of the plant's planned location in Sinop, on the Black Sea coast, cooling water is available at temperatures about 5 degrees below those at Akkuyu, which is located on the Mediterranean coast (see figure 2). Cooling water requirements of nuclear power plants exceed those of fossil-fueled power stations by about 25 percent, and the Sinop plant's 5-degree advantage would be expected to allow for about 1 percent greater output. Akkuyu had been under consideration as a site since the 1970s, while Sinop had been under consideration since 2006. As is evident from figure 2, Sinop would have been a much more practical and economical site for the Russian plant because it is much closer to Russia and would therefore reduce fuel transportation costs. Since the original fuel will come from Russia, and be transported back to Russia after usage for reprocessing and final storage, radioactive material will travel between the two countries. Hence, Sinop is obviously a much more attractive site for Russia. Although Russia has expressed interest in preparing a bid to build the nuclear plant at Sinop, it has been turned down by the Turkish side to avoid becoming too dependent on Russia. Currently, more than half of Turkey's natural gas needs are imported from Russia. With gas-fired power generation amounting to about half of the total supply, Russia would have had a dominant role in the Turkish electricity sector if it had been given the right to construct the second nuclear plant as well. Clearly, diversifying supply not only by technology but also by source country is a priority of Turkish policymakers. Goodwill agreements were signed with South Korea and Canada but were

FIGURE 2. Locations of the Akkuyu and Sinop Nuclear Power Plants

later canceled. The historical developments to determine which company would build the plant provide a better understanding of the events that led to Japan's selection.

Historical Developments

Since 2006 preparatory work has been under way to build a second nuclear plant in Sinop on the Black Sea coast, leading to goodwill agreements with three countries:

- In March 2010, a goodwill agreement was signed between Korea Electric Power Corp. and EÜAŞ to prepare a bid for constructing the plant at Sinop.

- In December 2010, a goodwill agreement was signed with Japan to prepare a bid for constructing the plant at Sinop.

- In April 2012, an agreement was signed between Canadian Candu Energy Inc. and EÜAŞ to conduct feasibility studies for the Sinop nuclear power plant project.

An agreement with South Korea could not be reached because KEPCO was not satisfied with the prospect of a guaranteed sales agreement with the electricity trading company TETAŞ and insisted on receiving a state-level guarantee for electricity sales during the operation phase.

Candu Energy Inc. submitted a feasibility report for the construction of a four-unit plant in Sinop, but the final decision of the Turkish government led to Japan's selection.[33]

Talks with Japan were suspended after the Fukushima accident in March 2011 but later resumed. In May 2013, an intergovernmental agreement was signed with Japan for "exclusive negotiating rights to build a nuclear power plant." It is a Japanese-French alliance led by Mitsubishi Heavy Industries and GDF Suez, which will operate the plant.

Resource Diversification

Another motive for awarding the second nuclear power plant project to a Japanese-led consortium is resource diversification away from Russia. As noted, Russia already has a dominant position as a primary energy supplier to the Turkish market, with 52.6 percent of Turkey's gas imports coming from Russia, implying that gas-fired power generation is dependent on Russian gas. Japan currently has no role in the Turkish electricity sector, either direct or indirect, so its selection will contribute to the diversification of resources by source country.

RISKS

The nuclear industry is surrounded by a variety of risks related to safety, operation, finance, and strategy. These risks can result from many sources: design/production processes, operation processes, training processes, social responsibility, public resistance, external influences (such as natural disasters, terrorist attacks, economic factors), financial processes, and so on. The risks that might threaten Turkey's nuclear program can be classified into five categories:

1. Political risks
2. Regulatory risks
3. Commercial and financial risks
4. Safety risks
5. Public risks

Political Risks

The AKP government is determined to proceed with its nuclear plans and achieve the goal of having two power plants in operation and a third one under construction by 2023. Opposition parties are criticizing the current government's plans for the development of nuclear power and have declared that they will seek a referendum. In March 2011, after the Fukushima accident, the deputy chairman of the main opposition party, CHP (Republican People's Party), urged the government to cancel the nuclear program and suggested that a regional referendum be carried out.[34] Similarly, the deputy chairman of the second-largest opposition party, MHP (National Action Party), has suggested a referendum on the issue.[35] Yet opposition parties are not categorically against nuclear power. Moreover, the MHP's party program explicitly refers to the development of nuclear energy generation technologies as a priority item under its energy policy chapter.[36] The CHP's party program adopts a more conservative viewpoint, saying that a safe storage solution of nuclear waste will be taken into account while the decision on nuclear power plant projects under consideration is being finalized.[37] The Akkuyu project should not be threatened by this as all nuclear waste is to be transported to Russia for reprocessing and/or final storage, but uncertainties remain for the second nuclear project, which will require Turkey to more fundamentally address the complex issue of waste disposal.

Geopolitical issues constitute another risk source in this category. The recent Ukraine crisis, for example, presents one of the most serious East-West confrontations since the Cold War and evolves with new sanctions on Russia.[38] As a non-EU country, Turkey is not bound by Brussels's decisions and the country's relationship with Russia can either improve or deteriorate depending on Turkey's position. It can become a risk for the Akkuyu project if the relationship deteriorates. According to Article 18 of the intergovernmental agreement between the two countries, either party may terminate the agreement at any time by way of one year's notice. Considering the fact that construction has not yet started, Russia can afford to terminate the agreement if the Ukrainian crisis escalates and Turkey takes a strict political position against Russia. Turkey, meanwhile, would be hard-pressed to attract a similarly beneficial nuclear power plant deal with another country.

Regulatory Risks

The existence of a transparent, well-established regulatory framework, together with an independent regulatory authority with well-defined regulations, licensing, and control procedures, is a prerequisite for healthy development of Turkey's nuclear energy plan. National safety requirements; site permit requirements; reactor licensing requirements; safety requirements; discharge authorizations; waste management, storage transport, and disposal requirements; and decommissioning requirements need all be clearly specified as part of the regulatory framework.

Turkey is still at the stage of establishing and updating its nuclear legislation. According to the nuclear power plant voluntary Exporters' Principles of Conduct,[39] before entering into a contract to supply a nuclear power plant to a customer, vendors should have made a reasonable judgment that the customer state has a legislative, regulatory, and organizational infrastructure either in place or under development for implementing a safe nuclear power program with due attention to safety. Moreover, the customer should have enacted national nuclear laws or developed a regulatory framework that formalizes and keeps current a credible national strategy and/or a plan to handle spent fuel and nuclear waste in a safe, secure, and environmentally sound manner. The Nuclear Power Plant and Reactor Exporters' Principles of Conduct can be considered as a potential risk for Turkey's second power plant since all the requirements are not yet fulfilled and both Mitsubishi Heavy Industries and GDF Suez, which constitute the consortium to construct the plant, are parties to the principles. The attachment of TAEK to the Ministry of Energy and Natural Resources,[40] which violates the principle of having an independent regulatory institution, and the incompleteness of Turkish regulations (not having a long-term strategy for nuclear-waste management) can be considered major weaknesses. Furthermore, licensing and regulatory risks may be triggered by findings from the Environmental Impact Assessment (so-called ÇED) report as well as from licensing procedures related to the Ministry of Energy and National Resources, the Energy Market Regulatory Authority, TAEK, and municipalities.

Commercial and Financial Risks

Since the agreed-upon sales price for the Akkuyu plant is independent of cost, all financial risk related to a possible cost overrun will need to be absorbed by the project company. Rosatom, the project company that will be constructing Turkey's first nuclear power plant, has argued that there could be a cost overrun by about 34 percent (from the initially planned cost of $18.7 billion to $25 billion). Rosatom is Russia's state atomic energy corporation, and its commitment and financial risk are thus indirectly Russia's commitment and risk. The Russian government's plan to sell some shares of Inter RAO (a diversified energy holding company managing assets in Russia, Europe, and Commonwealth of Independent States countries) to finance the project, announced in October 2013 by the Russian minister of energy, shows that the Russian government is ready to shoulder the financial risks. Alleviating the risk of a cost overrun for the realization of the project, the Russian government's possible subsidy opens up another potential risk: if there is any political conflict between Russia and Turkey, could it negatively affect the implementation of the Akkuyu project?

Safety Risks

The worst accident involving radioactivity since the 1986 explosion at Chernobyl occurred at the Fukushima Daiichi nuclear power plant following the March 2011 earthquake and tsunami. International reaction to the nuclear disaster has been diverse and widespread. Germany became the first industrialized power to agree on an end to nuclear power following the disaster in Japan. Germany closed all of its old nuclear reactors immediately and decided to phase out the rest by 2022.[41] In Italy there was a referendum in which 94 percent voted against the government's plan to build new nuclear power plants.[42] The Swiss government also recently decided to abandon plans to build new nuclear reactors.[43] The British government, meanwhile, is going forward with nuclear plans and is set to sign a deal with France's EDF to start construction of the first post-Fukushima nuclear plant in Europe.[44] The Turkish government's nuclear agenda has not been affected by the Fukushima accident, though public opposition to nuclear power has grown. It would grow further should there be another nuclear accident, and that would increase the risk to the realization of Turkey's nuclear program.

Public Risks

Public demonstrations against nuclear power have been increasing in the past few years in Akkuyu and Sinop as well as in other major Turkish cities.[45] To preempt more acute demonstrations, the government should enhance the transparency of the nuclear power program and implement a more effective public communication strategy. The lack of such a strategy can be considered as a public risk that could lead to a delay in Turkey's nuclear program.

CONCLUSION

Turkey's motivation for nuclear power is triggered by the fact that the technology has been considered by policymakers to be a cheap, clean, and reliable source of energy generation. Turkish policymakers have been struggling to ensure continuous investment in electricity generation to meet the country's rapidly growing demand, which has averaged in excess of 7 percent per year in the past decade. The trend is expected to continue so that even conservative projections forecast a doubling of electricity demand within ten years. Reserves of domestic fossil fuels that could be used for power generation are negligibly small, with the exception of lignite, whose pollutant emissions lead to environmental concerns. While emissions of sulfur dioxide, nitric oxide, nitrogen dioxide, and other pollutants and particulates have been creating concern at the local level, emissions of carbon dioxide are creating concern at the global level. With an increase of 115 percent during 1990–2010, Turkey had the fastest growth rate of carbon emissions among Annex 1 countries party to the United Nations Framework Convention on Climate Change. The country's emission intensity needs to be reduced through an expansion of clean energy sources. The potential of renewable energy sources for power generation is large compared with most European countries, but the energy sources are bound by technical constraints on grid integration. As a zero-carbon electricity-generating source, nuclear power emerges as a favorable option for policymakers to ensure security of supply under grid conditions where a sufficient amount of renewables cannot be accommodated.

While being attractive in terms of carbon emissions, nuclear power also appears to be advantageous for Turkey from an economic viewpoint: the agreed-upon sales price for nuclear power from Akkuyu, where the country's first nuclear power plant will be constructed, is in real terms significantly lower than today's average electricity wholesale price in Turkey. Furthermore, according to the intergovernmental agreement with Russia for the construction of a nuclear power plant at the Akkuyu site, there is no financial risk on the Turkish side. The agreement is also advantageous for Turkey in terms of waste management as the spent fuel will be transported back to Russia for reprocessing or final disposal.

However, despite all these facts in support of policymakers' decision to proceed with nuclear power, various risks and concerns are associated with Turkey's nuclear move. Political, regulatory, and safety/public risks surround the Russian agreement, and public protests of nuclear power have been on the rise in Turkey.

To ensure a smooth introduction to nuclear power that addresses the risks and concerns, a transparent process needs to be implemented along with a healthy institutional system enabling public trust and support for nuclear energy generation. More specifically, in the case of Turkey, the emerging major recommendations are as follows:

1. A professional public relations campaign should be implemented, communicating all the details of the nuclear program in general and individual projects in particular with emphasis on safety, security, and nonproliferation measures, to ensure public trust and support.

2. Strict control of all safety and security measures, both during construction and subsequent operational phases, should be carried out in a transparent manner by an independent regulatory institution.

3. An independent regulatory institution should be established, or the Turkish Atomic Energy Authority should be decoupled from the Ministry of Energy and Natural Resources and restructured as an independent entity.

4. Domestic regulations should be revised to be precise and clear on all aspects and to impose a long-term strategy for safety, security, and waste management issues.

For the construction of Turkey's second nuclear power plant in Sinop on the Black Sea coast, in spite of Russia's expressed interest, Japan has been chosen as a partner in order to diversify ownership. In contrast to the Akkuyu agreement, two issues may be challenging for the Turkish side in Sinop.

1. Economics: estimated construction costs have already escalated for the Akkuyu project, but all the financial risk is on the Russian side. For the Sinop project, the Turkish government, through its 30 percent partnership of EÜAŞ, will take on a significant part of the financial risk. As for waste management, spent fuel from Akkuyu will be transported back to Russia for reprocessing or final disposal. But for the Sinop project, the government has to come up with a different solution that must take into account the need for a long-term nuclear-waste-disposal strategy.

2. Technical issues: another potential problem for the Sinop plant may occur due to technological differences between the Russian and Japanese plants. Such differences would hinder the duplication of regulatory practices from Akkuyu and necessitate the development of unique technology-specific standards, rules, and regulations for the licensing and operation of the Sinop plant.

If these issues are not resolved as satisfactorily as in Akkuyu, public trust and support for the Sinop plant may be more difficult to ensure.

NOTES

1   Mustafa Kibaroğlu, "Turkey's Quest for Peaceful Nuclear Power," *Nonproliferation Review* (Spring-Summer 1997).

2   M. Özcan Ültanır, *21. Yüzyıla Girerken Türkiye'nin Enerji Stratejisinin Değerlendirilmesi* (Evaluation of Turkey's 21st-Century Energy Strategy) (Istanbul: TÜSİAD, December 1998).

3   H. Palabıyık and H. Yavaş, "Başlamayan Senfoni: Türkiye'nin Nükleer Santral Serüveninin Üzerine" (Starting Symphony: The Adventure of Turkey's Nuclear Power Plant Is Over), Çanakkale Onsekiz Mart University, *Journal of Management Science* 4, no. 2 (2006): 17–25.

4   Sinan Ülgen and Aaron Stein, "Efforts to Control the Atom and the Transfer of Nuclear Technology: An Evaluation from Turkey's Perspective," *The Turkish Model for Transition to Nuclear Energy – II* (Istanbul: Centre for Economics and Foreign Policy Studies, December 2012).

5   Ültanır, *Evaluation of Turkey's 21st-Century Energy Strategy.*

6   Kibaroğlu, "Turkey's Quest for Peaceful Nuclear Power."

7   Ibid.

8   For a more detailed analysis of the issue, see Kibaroğlu, "Turkey's Quest for Peaceful Nuclear Power," 38–39.

9   Erkan Erdogdu, "Nuclear Power in Open Energy Markets: A Case Study of Turkey," *Energy Policy* 35, no. 5 (May 2007).

10 Turkish Statistical Institute, "Power Installed of Power Plants, Gross Generation and Net Consumption of Electricity," www.turkstat.gov.tr/PreTablo.do?alt_id= 1029.

11 Ültanır, *Evaluation of Turkey's 21st-Century Energy Strategy.*

12 Ibid.

13 World Nuclear Association, "Nuclear Power in Turkey," updated May 2014, http://world-nuclear.org/info/Country-Profiles/Countries-T-Z/Turkey/#.UkzdA4bIV27.

14 Aaron Stein, "Turkey's Nuclear Timeline," *Nuclear Wonk: Nuclear and Political Musings in Turkey and Beyond* (blog), http://turkeywonk.wordpress.com/2012/12/03/turkeys-nuclear-timeline.

15   Turkish Electricity Trading and Contracting Company, Inc. (TETAŞ), "2012 Activity Report."

16   World Nuclear Association, "Nuclear Power in Turkey."

17   "Russia to Sell Inter RAO Shares to Finance Akkuyu Plant," *Today's Zaman*, October 10, 2013, www.todayszaman.com/news-328696-russia-to-sell-inter-rao-shares-to-finance-akkuyu-plant.html.

18   World Nuclear Association, "Nuclear Power in Turkey."

19   Ibid.

20   T.C. Enerji ve Tabii Kaynaklar Bakanlığı, "Sinop Milletvekili Sayın Engin Altay'ın Yazılı Soru Önergesi ve Cevapları (Question and Answer Session With Sinop Deputy Engin Altay), July 29, 2013, Türkiye Büyük Millet Meclisi Başkanlığı (Grand National Assembly of Turkey), http://www2.tbmm.gov.tr/d24/7/7-25371c.pdf.

21   "Sinop'taki nükleer santralin son iki ünitesini Türk mühendisler yapacak" (Turkish Engineers Are Working on the Last Two Units of the Nuclear Power Plant in Sinop), *Zaman*, May 7, 2013.

22   "3. nükleer santral için ilk tercih İğneada" (İğneada Is the First Choice for Nuclear Power Plants), *Zaman*, November 30, 2012.

23   "Yıldız, 3. nükleer santralin yeri henüz tespit edilmiş değil" (Yıldız: Nuclear Power Plant Location Has Not Yet Been Identified), Bloomberg HT, October 3, 2013, www.bloomberght.com/haberler/haber/1436671-yildiz-3-nukleer-santralin-yeri-henuz-tespit-edilmis-degil.

24   Turkish Ministry of Foreign Affairs, "Turkey's Energy Strategy," www.mfa.gov.tr/turkeys-energy-strategy.en.mfa.

25   International Atomic Energy Agency, "IAEA Support for Turkey's Nuclear Power Programme," November 14, 2012, www.iaea.org/NuclearPower/News/2012/2012-11-14-INIG.html.

26   "Yıldız: Nükleer santral 3,500 istihdam yaratıyor" (Yıldız: Nuclear Power Plant Creates 3,500 Jobs), Bloomberg HT, January 21, 2013.

27   Turkish Ministry of Foreign Affairs, "Turkey's Energy Strategy."

28   *2013–2022 Yılları Türkiye İletim Sistemi Bölgesel Talep Tahmin ve Şebeke Analiz Çalışması: Metodoloji ve Özet Sonuçlar* (Turkey Regional Transmission System Network Analysis and Demand Forecast Study: Methodology and Summary Results) (Ankara: Tübitak and Teiaş, June 21, 2013), www.teias.gov.tr/Dosyalar/TürkiyeBölgeselTalepTahminPlaniv3.pdf.

29   International Energy Agency, "Key World Energy Statistics 2013."

30   Assuming that retired capacity will be replaced by the same technology—or else, if not only additional but also retired thermal power generation capacity is replaced by zero carbon technologies, emissions would decline. A slight upward effect on emissions would still be present arising from the worsening of generation efficiency of thermal power plants over time.

31   Sinan Ülgen, ed., *The Turkish Model for Transition to Nuclear Power* (Istanbul: EDAM, 2011).

32   Ibid.

33 In March 2013 Candu Energy Inc. announced that, contrary to some media reports, it had not pulled out of a bid for the construction of a nuclear power plant in Sinop.

34 Zafer Çakmak, "CHP: Nükleer programdan vazgeçin" (CHP: Give Up Nuclear Program), IHA, March 17, 2011, www.iha.com.tr/gundem/chp-nukleer-programdan-vazgecin/165088.

35 "MHP nükleer için referandum istedi" (MHP Asked for a Referendum on Nuclear Power), CNN Turk, March 18, 2011, www.cnnturk.com/2011/turkiye/03/18/mhp.nukleer.icin.referandum.istedi/610432.0.

36 "MHP Parti Programı"(MHP Party Program), November 8, 2009, www.mhp.org.tr/usr_img/_mhp2007/kitaplar/mhp_parti_programi_2009_opt.pdf.

37 "CHP Parti Programı" (CHP Party Program), www.chp.org.tr/wp-content/uploads/chpprogram.pdf.

38 David M. Herszenhorn, Michael R. Gordon, and Alissa J. Rubin, "U.S. Announces New Sanctions in Ukraine Crisis," *New York Times*, March 6, 2014, www.nytimes.com/2014/03/07/world/europe/ukraine-sanctions.html?hpw&rref=world.

39 "Nuclear Power Plant Exporters' Principles of Conduct," January 16, 2013, http://nuclearprinciples.org/wp-content/uploads/2013/01/Principles_of_Conduct_TOKYO_JANUARY_2013.pdf.

40 See the Turkish Ministry of Energy and Natural Resources website, www.enerji.gov.tr/index.php?dil=en&sf=webpages&b=relatedcorporations&bn=1076&hn=&id=4812.

41 Deborah Cole, "Fukushima Fallout: Germany Abandons Nuclear Energy," *Sydney Morning Herald*, May 31, 2011, www.smh.com.au/world/fukushima-fallout-germany-abandons-nuclear-energy-20110530-1fczb.html.

42 "Italy Nuclear: Berlusconi Accepts Referendum Blow," BBC News, June 14, 2011, www.bbc.co.uk/news/world-europe-13741105.

43 James Kanter, "Switzerland Decides on Nuclear Phase-Out," *New York Times*, May 25, 2011, www.nytimes.com/2011/05/26/business/global/26nuclear.html?_r=0.

44 "Britain to Build Europe's First Nuclear Plant Since Fukushima," Reuters, October 20, 2013, http://uk.reuters.com/article/2013/10/20/uk-britain-nuclear-hinkley-idUKBRE99J03X20131020.

45 "Akkuyu'da nükleer karşıtı miting Mersin Akkuyu'ya kurulması planlanan nükleer santral protesto edildi" (Anti-Nuclear Protest Rally Against Akkuyu Nuclear Power Plant Planned for Mersin), NTV/MSNBC, August 10, 2009, www.ntvmsnbc.com/id/24990457; "Kadıköy'de 'Nükleer Karşıtı' Eylem" ("Anti-Nuclear" Actions in Kadikoy), Haberler.com, April 24, 2011, www.haberler.com/kadikoy-de-nukleer-karsiti-eylem-2677215-haberi; "Sinop'ta nükleer karşıtı eylem" (Anti-Nuclear Actions in Sinop), Haberler.com, November 2, 2013, www.haber1.com/sinopta-nukleer-kars-iti-eylem.html.

CHAPTER 2

REGULATING NUCLEAR POWER
The Case of Turkey

İZAK ATIYAS

INTRODUCTION

Ensuring the safety of nuclear energy is of paramount importance as Turkey embarks on adopting a nuclear energy program. The key to nuclear safety is having in place an effective regulatory framework. It is useful to think of regulatory frameworks for nuclear safety as consisting of two main components. The first entails international agreements and rules, regulations, standards, and guidelines established by international organizations active in the nuclear power area. Becoming a party to international agreements or conventions, or, in the case of the European Union (EU), adoption of EU directives, therefore stands out as an important component of the regulatory arsenal. The second component consists of the regulatory apparatus established within countries. This regulatory apparatus also entails policies, rules, and regulations as well as enforcement mechanisms, some of which serve the purpose of carrying out obligations spelled out in international or transnational rules. As a result, the extent to which domestic laws and regulations actually capture and reflect the spirit of international norms is an important dimension of the quality of the domestic regulatory framework. Perhaps more important in the domestic context, international norms and experience show that establishing a regulatory authority and ensuring its independence from both the government and nuclear energy operators are crucial elements of this apparatus. Another important element is instituting mechanisms that guarantee the transparency and accountability of the regulatory framework. So the institutional and governance characteristics of the domestic regulatory framework are also important determinants of its quality.

To appreciate the importance of the regulatory framework, and especially the independence of the regulatory authority, it is useful to recall a crucial characteristic of nuclear energy, namely risk. Nuclear energy entails a multitude of safety and financial risks. Moreover, these risks are interrelated. Measures or regulations that aim at reducing safety risks to an acceptable minimum would unavoidably increase the costs of building and operating nuclear plants or managing spent fuel or radioactive waste, thereby decreasing the profitability of nuclear energy. Under some circumstances, implementation of safety measures may even require that generation of electricity be suspended. Hence the interests of nuclear plant operators or even ministries responsible for the supply of electricity may be in direct conflict with rules and regulations that ensure safety. Consequently, ensuring safety requires that decisions are made independently, and in some circumstances against the interests, of nuclear plant operators or the ministry responsible for energy. The independence of the regulatory authority from the government and operators is therefore increasingly seen as a critical element of the regulatory framework.

This increased concern with the institutional qualities of regulation, such as independence and transparency, reflects the appreciation that institutional weaknesses have played an important role in nuclear accidents. The Fukushima incident is highly illuminating in that respect. The main conclusions of the report of the Japanese parliament on the Fukushima accident highlight the failures of the institutional setup:

> The TEPCO [Tokyo Electric Power Co.] Fukushima Nuclear
> Power Plant accident was the result of collusion between the
> government, the regulators and TEPCO, and the lack of
> governance by said parties. They effectively betrayed the
> nation's right to be safe from nuclear accidents. Therefore, we
> conclude that the accident was clearly "manmade." We
> believe that the root causes were the organizational and regu-
> latory systems that supported faulty rationales for decisions
> and actions, rather than issues relating to the competency of
> any specific individual.[1]

The report states that the direct causes of the accident were foreseeable but that the parties involved, namely the operator, the regulatory bodies, and the government body promoting the nuclear power industry (the Ministry of Economy, Trade, and Industry), "failed to correctly develop the most basic safety requirements."[2] These findings highlight the lack of independence of the regulatory authority and a high degree of "capture."

Conformity of primary and/or secondary legislation with international norms—assuming that such norms represent an element of best practice—is a desirable feature of domestic regulatory frameworks. But this is not sufficient in itself to ensure that the domestic regulatory framework reaches a high degree of quality. The quality of the regulatory framework also depends on how these regulations are implemented. Here again the literature emphasizes the independence of the regulatory authority as an important determinant of the quality of the regulatory framework. In particular, the literature on independent regulatory authorities emphasizes that there may be significant gaps between the de jure and de facto independence of these regulatory authorities. In many countries, while laws may declare regulatory authorities to be independent, in practice their decisions may be subject to pressure from the government or operators or both. It is also generally accepted that the likelihood of independence is enhanced by adoption of some internationally accepted conditions.

Turkey is no stranger to independent regulatory authorities. The Capital Markets Board, the first independent regulatory authority in Turkey, was established as early as 1982. In the 1990s, and especially in the 2000s, others have been established in such disparate areas or sectors as competition, banking, electronic communications, energy, public tenders, tobacco, and sugar. As Turkey is a candidate for European Union accession, many of these initiatives have been inspired by EU legislation and directives. For example, Turkish competition law is designed after the relevant articles of the Treaty on the Functioning of the European Union, and its implementation follows closely decisions by the European Commission and the European Court of Justice. Similarly, regulations in the electronic communications or energy industries closely follow the relevant EU directives. While the degree of actual independence of these independent regulatory authorities varies, their establishment has played an important role in the creation of a relatively nondiscriminatory and competitive business environment.[3]

The discussion that follows will provide an overview of the international and EU norms on nuclear safety and regulation as well as the independence of regulatory authorities in the nuclear energy area, summarize lessons learned from international experience about independence and transparency of regulatory authorities, and evaluate the Turkish case.

INTERNATIONAL NORMS FOR NUCLEAR ENERGY

The development of an international legal framework in the area of nuclear energy has a long history. The International Commission on Radiological Protection was established in 1928. The International Atomic Energy Agency, the European Atomic Energy Community, and the Organization for Economic Cooperation and Development's Nuclear Energy Agency (OECD-NEA) were established in the 1950s. It was not until after the Chernobyl accident in 1986, however, that the will to increase international cooperation was greatly enhanced. Countries became more willing to engage in international cooperation in areas that until then were considered to be under sovereign jurisdiction.[4]

The Convention on Nuclear Safety

A milestone in these efforts was reached with the adoption of the Convention on Nuclear Safety in 1994. The CNS is a legally binding (though, as we'll discuss, not truly enforceable) document that stipulates a number of fundamental safety principles for nuclear power plants that are to be followed by the contracting parties. More detailed safety standards are left to internationally formulated safety guidelines that are "updated from time to time and so can provide guidance on contemporary means of achieving a high level of safety" (Preamble, viii).

The fundamental safety principles elucidated in the CNS cover a wide range of areas such as the regulatory framework, the regulatory body, responsibilities of the license holder, financial and human resources, quality assurance, emergency preparedness, and design and construction. On the specific issue of the regulatory framework, the CNS stipulates that each contracting party will establish a legislative and regulatory framework that entails national safety regulations and requirements, a system of licensing, a system of inspection for compliance, and enforcement mechanisms including license suspension, modification, and revocation (Article 7). Article 8 of the CNS requires each contracting body to establish a regulatory body entrusted with the implementation of these regulations and to ensure an effective separation between the functions of the regulatory body and those of any other body or organization concerned with the promotion or utilization of nuclear energy. Thus, the notion of independence established in the CNS is limited to the separation of the regulatory body from promotional and operational functions. The CNS does not entail a broader view of independence.

As emphasized in the literature,[5] in the CNS the requirement to establish a legal and regulatory framework is quite general and does not stipulate detailed elements. While the CNS is supposed to be binding, it lacks an enforcement mechanism or a system of sanctions that can be applied in the event of noncompliance. For effectiveness, it depends on a process of peer review and peer pressure, recognizing that all operators of nuclear power plants share an interest in preventing an accident anywhere, as that would harm the credibility of the nuclear industry everywhere.[6] The CNS stipulates that contracting parties prepare a report on measures taken to meet the nuclear safety requirements specified in the convention. The reports are to be prepared prior to meetings, which are to be held every three years, and presented at the meetings to other contracting parties. During the meetings, other contracting parties can ask questions about the reports or present comments on them. Contracting parties in turn are encouraged to implement the improvements recommended during the meetings.

Selma Kuş lists four weaknesses in the peer review mechanism.[7] First, there are no sanctions if a country does not submit a report or send representatives to the meetings. Second, the completeness and accuracy of the reports are at the discretion of the submitting country. In other words, there are no mechanisms to verify the accuracy of the reports. Third, the review process emphasizes the formal and procedural obligations of the convention rather than substantial implementation. Fourth, the review process does not allow the review of the safety of individual nuclear installations. These shortcomings are especially important given that there is often a significant gap between the formal characteristics of regulation and their implementation in specific country environments, an issue further discussed below.

IAEA Standards

The most important vehicle for setting international standards in the area of nuclear safety is the International Atomic Energy Agency. In 2006 the IAEA adopted the Fundamental Safety Principles.[8] The ten fundamental safety principles adopted in this document are as follows:

1. The primary responsibility for safety must rest with the person or organization charged with facilities and activities that give rise to radiation risks.

2. An effective legal and governmental framework for safety, including an independent regulatory body, must be established and sustained.

3. Effective leadership and management for safety must be established and sustained in organizations concerned with, and facilities and activities that give rise to, radiation risks.

4. Facilities and activities that give rise to radiation risks must yield an overall benefit.

5. Protection must be optimized to provide the highest level of safety that can reasonably be achieved.

6. Measures for controlling radiation risks must ensure that no individual bears an unacceptable risk of harm.

7. People and the environment, present and future, must be protected against radiation risks.

8. All practical efforts must be made to prevent and mitigate nuclear or radiation accidents.

9. Arrangements must be made for emergency preparation and response to nuclear or radiation incidents.

10. Protective actions to reduce existing or unregulated radiation risks must be justified and optimized.

These principles are treated as "soft law," that is, they are recommendations only, not binding on members of the IAEA.[9]

On the basis of these Fundamental Safety Principles, which apply for all nuclear facilities and activities, the IAEA has identified seven "General Safety Requirements," which also apply to all facilities and activities:

1. Governmental, legal, and regulatory framework for safety

2. Leadership and management for safety

3. Radiation protection and safety of radiation sources

4. Safety assessment for facilities and activities

5. Predisposal management of radioactive waste

6. Decommissioning and termination of practices

7. Emergency preparedness and response

The IAEA is publishing General Safety Guides on how these requirements may be met.

The IAEA also has developed guidelines on what needs to be done by countries that decide to develop nuclear energy. This fundamental approach, called "Milestones in the Development of a National Infrastructure for Nuclear Power,"[10] aims to reveal in a systematic manner the steps that are required to be taken by countries that plan to establish a nuclear power plant. The study is based on the acknowledgement that entering the field of nuclear energy is an extremely complicated process. The "infrastructure" encompasses a number of elements ranging from the choice of location, physical facilities, and equipment to the relevant legal and regulatory framework. The study is focused especially on the planning, tender preparation, construction, and commissioning phases. However, the operation, radioactive waste management, and decommissioning phases also have been considered to the extent that they are necessary in the initial plan. According to the study, issues relating to phases such as operation, spent fuel and waste management, and decommissioning also should be considered at the phase of participation in the tender, and the planning process should encompass these requirements.

FEATURES OF THE REGULATORY AUTHORITY ACCORDING TO THE IAEA

Finally, it would be beneficial to summarize how the IAEA conceptualizes the independence of the regulatory authority. As mentioned, establishment of an independent regulatory body is identified as one of the IAEA's ten fundamental safety principles. The desirable qualities of the regulatory body are further explained in the General Safety Requirement concerning the Governmental, Legal, and Regulatory Framework for Safety.[11] This document lists the functions and responsibilities of the government (organized under fourteen "requirements") and those of the regulatory body (21 requirements). It also stipulates rules governing responsibilities regarding international obligations as well as international cooperation and exchange of information. The governmental responsibilities include the establishment of a national policy and strategy for safety, a legal framework for safety, and a regulatory body, along with maintaining the latter's independence.

Regarding independence, the document acknowledges that an independent regulatory body will not be "entirely separate from other governmental bodies." Nevertheless, it states that "the government shall ensure that the regulatory body is able to make decisions under its statutory obligation for the regulatory control of facilities and activities, and that it is able to perform its functions without undue pressure or constraint." The document specifies three key conditions for "effective independence": sufficient authority, staffing, and financial resources.[12] Authority to collect information and authority to make inspections are highlighted.

The document also prescribes a number of measures regarding the accountability and transparency of the regulatory process. It states that "the government or the regulatory body shall establish, within the legal framework, processes for establishing or adopting, promoting and amending regulations and guides."[13] In addition, the process will entail "consultation with interested parties" and the regulatory body will "notify interested parties and the public" of the principles and associated criteria for safety established in its regulations.[14] The regulatory body will also establish "appropriate means of informing and consulting interested parties and the public about the possible radiation risks associated with facilities and activities."[15] The regulatory body is required to "hold meetings to inform interested parties and the public." In particular, it is to consult with anyone living near the facility through an "open and inclusive process."[16]

Many of these principles are restated and further explained in a report, *Independence in Regulatory Decision Making*,[17] prepared by the International Nuclear Safety Advisory Group.[18] In particular, the report provides details about measures related to accountability and transparency. First of all, the report explicitly states that "appropriate procedures that are open to public scrutiny for eliciting opinions from licensees and other stakeholders, and procedures" may be used as "legal barriers to protect the independence in regulatory decision making from external interference in decisions on specific safety issues."[19] Second, while discussing the regulatory body's decisionmaking process, the report suggests that all relevant information used during the process, including opinions elicited from external sources, should be documented.[20] Furthermore, in addition to the decision, the regulatory body's legal and technical justification needs to be properly documented, "that is, the regulatory evaluation of the information used as a basis for the decision." In addition, the report mentions transparency not only to the regulated entities but also to the public, as a key measure to ensure independence. Hence it states that the regulatory body should have the authority as well as an obligation to communicate regulatory decisions and the underpinning documentation to the public:

> By means of such public access to information, the independence in regulatory decision making can be open to public scrutiny. At the same time, this serves to fulfill the requirement

for the regulatory body to be accountable to the public, whose health and safety it is responsible for protecting.[21]

The International Nuclear Safety Advisory Group report is noteworthy for two reasons. First, it is one of the rare cases of official recognition of the significance of a requirement that justifications behind regulatory decisions, including the information that led to the decision, should be transparent. Second, it establishes a clear link between transparency and the degree of independence of the regulatory agency.

REGULATORY FRAMEWORK FOR NUCLEAR SAFETY IN THE EUROPEAN UNION

In the European Union, the safety directive forming the "community framework for the nuclear safety of nuclear installations" was adopted in 2009. The directive is based on CNS and IAEA standards. The important difference is that it is binding and enforceable. In particular, the European Commission may initiate an infringement procedure if a member state fails to comply, or it may refer the member state to the European Court of Justice, as per the Treaty on the Functioning of the European Union.

Similar to the CNS, the safety directive aims to provide a set of minimum standards regarding nuclear safety, rather than presenting a specific legal and regulatory framework. According to Article 4 of the directive, each member state is obliged to establish a national legal, regulatory, and organizational framework with regard to nuclear safety. Article 5 stipulates that the regulatory authority should be "functionally separate from any other body or organisation concerned with the promotion, or utilisation of nuclear energy, including electricity production, in order to ensure effective independence from undue influence in its regulatory decision making." Member states are obliged to submit a report analyzing the degree of enforcement of the safety directive until 2014, and then once every three years. In contrast to the CNS, failure to submit the report could result in sanctions at the EU level. Furthermore, member states are obliged to conduct a self-assessment and to subject it to a peer review once every ten years.

According to Ana Stanic, there are two main deficiencies in the safety directive.[22] First, it does not envisage surprise inspections of power plants and independent verifications can be carried out only by national authorities, not those from the EU. Second, it does not require that reports prepared under the directive be made public. Indeed, Article 8 of the directive, which addresses the topic of "information to the public," is quite generic, merely requiring member states to "ensure that information in relation to the regulation of nuclear safety is made available to the workers and the general public." These requirements are weaker than those in the International Nuclear Safety Advisory Group report.

After the Fukushima incident, the European Commission prepared a draft proposal for a safety directive amending the Nuclear Safety Directive. One of the objectives of the proposed amendments is to strengthen the role and effective independence of national regulatory authorities. The proposed changes in the area of transparency are noteworthy. The competent regulatory authority and the license holder would be required to develop a transparency strategy, "which covers information provision under normal operating conditions of nuclear installations as well as communication in case of accident or abnormal event conditions." The role of the public is expanded through the requirement that it effectively participate in the licensing process of nuclear installations. Specifically, the proposal stipulates that "Member States shall ensure that the public shall be given early and effective opportunities to participate in the licensing process of nuclear installations, in accordance with relevant Union and national legislation and international obligations." At the same time, the proposed directive in some respects falls short of the recommendations in the International Nuclear Safety Advisory Group report. In particular, the directive does not require that the justification for regulatory decisions be documented as well as the decisions themselves, or that the regulatory authorities be obligated to establish public access to such documentation.

Besides the CNS, the second landmark in the development of international law on nuclear safety was the adoption of the Joint Convention on the Safety of Spent Fuel Management and the Safety of Radioactive Waste in 1997. The joint convention requires the contracting parties to take appropriate legislative, regulatory, and administrative measures to govern the safety of spent fuel and radioactive waste management and to ensure that individuals, society, and the environment are adequately protected against radiological and other hazards by regulating the appropriate siting, design, and construction of nuclear facilities. The joint convention also stipulates a reporting and peer review mechanism for effectiveness. The reports are required to include lists of spent fuel and radioactive waste management facilities and inventories of spent fuel and radioactive waste. As was the case with the CNS, however, the joint convention does not entail an effective enforcement mechanism or sanctions in case of noncompliance.

LESSONS LEARNED

The need for regulation, and therefore the problem of ensuring high-quality regulation, arises not only in nuclear energy but also in the activities of many different sectors.

It will be useful to remember why regulation is needed in the first place. As mentioned in the introduction, nuclear energy poses a number of inter-related safety and financial risks. The problem is that these risks are internalized by operators only to a limited extent. In the event of an accident, a large part of the cost is borne not by the operators, or even by the government, but by society. This cost constitutes negative externalities on society and cannot be properly managed by the market mechanism. That is why nuclear risks and the problem of safety in particular call for a mechanism that regulates the behavior of operators and other stakeholders.

The distinguishing feature of the problem of nuclear safety is that the probability of an accident is relatively low, and the cost in case an accident occurs is extremely high. Beyond this distinguishing feature, however, the fundamental problem of externalities, and more generally of market failure in nuclear energy, is not different from similar problems observed in other sectors or activities where these problems exist. The hazard of market failures has become prevalent in the past two to three decades: globalization, liberalization, and privatization have generally increased the role of the market mechanism in economic activities relative to earlier periods when governments took a more direct and active role. The withdrawal of the state from economic activities has created a new role for the state, one of regulating the behavior of market players so as to reduce social harms associated with market failures.

The need for government intervention to correct for market failures has raised an important question: how to make sure that this authority to intervene is used in the public interest and is not influenced by the interests of politicians or of market players—the very firms whose behavior needs to be regulated. In other words, how can "capture" be avoided? The answer to that question involves delegating regulatory authority to special agencies—independent regulatory authorities—that are (or are rendered) insulated from the influence of politics and of market players. The independence of an independent regulatory authority is not absolute: its authority is limited by the constituting laws. In fact, the generally accepted division of labor is that policy is the responsibility of the government, and the creation and enforcement of the regulatory framework that is necessary to implement the policy is the task of the independent regulatory authority. In this context, its independence has become a crucial determinant of the quality of the regulatory framework.

It was indicated above that one of the important internationally accepted conditions for nuclear safety is to ensure the independence of regulatory authorities from political influence. The literature on independent regulatory authorities suggests a number of specific measures to ensure that their independence goes beyond those identified in conventions, IAEA norms, or EU directives. The first of such measures is that the government should not have the authority to overturn the decisions of the independent regulatory authority. The purpose of this condition is self-explanatory and plays a crucial role in ensuring that the regulatory authority's decisions and the regulations it puts out are immune from political influence. The second requirement is that decisionmakers in the regulatory authority, namely the head of the agency and/or members of the governing board, should be appointed for fixed turns, the duration of which often exceeds the regular terms of governments. Moreover, those who have been appointed should be protected from job termination in the sense that they can be removed from their duties only under extraordinary circumstances such as in the event of some wrongdoing or health problems. The purpose of this condition is to protect regulators from threats of being removed for not following the wishes of political actors or commercial interests that may be contrary to safety regulations. This way, it is hoped that appointments will be based on merit and professionalism rather than political affinity. The third condition is that the agency should have financial independence; this independence may be ensured by allowing the agency to have access to financial resources not directly controlled by the government, such as earmarked taxes.

Independence should not imply absence of accountability. One important measure of accountability is that decisions by an independent regulatory authority should be subject to judicial oversight, such as through an appeal mechanism. In Turkey, for most independent regulatory authorities, this appeal body is the Council of State, the high court of appeals for administrative decisions. Second, the regulatory authority, despite being independent, should be subject to reporting requirements and perhaps performance evaluation. For example, the performance of the independent regulatory authorities can be evaluated in parliamentary hearings.

Transparency is an important determinant of the quality of regulation, and indeed of the degree of independence that an independent regulatory authority actually exercises. The scope of the principle of transparency can be quite wide. Ensuring public access to regulatory decisions and publishing them in the official gazette and on the authority's website are minimum requirements of transparency. Transparency should be further enhanced by making public the justification behind decisions, as well as technical studies that played a role in reaching the decisions. Another transparency measure is the process of public consultation, that is, publishing drafts of regulations and decisions and soliciting comments from interested parties and citizens in general. In Turkey, public consultations regarding secondary legislation have become a standard executed by most regulatory authorities. A further step could be to make public the comments received during the consultation process.

How does transparency affect the quality of regulation and, in particular, the independence of the regulatory authority? Transparency allows scrutiny by the media, nongovernmental organizations, trade unions, professional associations, the legal community, and academia by enabling review and criticism as well as a chance to propose alternatives. Furthermore, transparency allows stakeholders increased opportunity to launch challenges through judicial review of regulations and decisions they deem unlawful. The prospect of such criticism and review provides incentives to regulatory authorities to take their business seriously in the first place.

Hence independence and transparency are necessary to endow regulatory agencies with the correct incentives. Another crucial determinant of regulatory quality is technical capacity or more generally the level of human capital in the independent regulatory authority. The availability of the necessary skills depends on the quality of education and training in the country as well as the human resources policy of the regulatory authority. The existence of degree programs in nuclear and related engineering sciences in the country increases the availability of necessary skills. Of course, even technical capacity has an incentive dimension, because human resource policy also is influenced by the degree to which the independent regulatory authority's recruitment efforts are based on merit. Assuming merit-based recruitment, over the medium term, one should expect that technical capacity and skills act as a relatively less binding constraint, especially if the nuclear energy development program in the country sets as a target the development of the necessary skills. In other words, while to some extent technical capacity is a constraint on regulatory quality, it is also partly a consequence of the latter over the medium term. A regulatory authority that is truly independent and that is keen on high regulatory quality should not face serious problems in enhancing technical capacity over time.

AN EVALUATION OF THE REGULATORY FRAMEWORK FOR NUCLEAR ENERGY IN TURKEY

The Turkish Atomic Energy Authority (TAEK), discussed in detail below, is the designated regulatory authority for nuclear safety and security. Two ministries take an active role in nuclear power plant projects: the Ministry of Energy and Natural Resources is responsible for energy policy and ultimately for the security of Turkey's electricity supply, and the Ministry of Environment and Urbanization is responsible for the protection of the environment. It implements the Decree on Environmental Impact Assessment (1997), which requires operators of facilities, including of nuclear power plants, to prepare environmental impact assessment reports during a project's planning stage. These reports are submitted to the Ministry of Environment and Urbanization, and construction cannot proceed until the ministry approves the projects.

The legal framework of nuclear energy in Turkey is based on two laws: the Law on the Construction and Operation of Nuclear Power Plants and Energy Sale (Law No. 5710, known as the "Nuclear Law"), which was enacted in 2007, and the Law on TAEK (Law No. 2690), enacted in 1982. TAEK is the authority responsible for nuclear safety and security, and the Nuclear Law also stipulates that until a new agency with the task of regulation and supervision of nuclear activities is established, TAEK will serve as the regulatory authority. Law No. 2690 authorizes TAEK to ensure nuclear safety and security by licensing and inspecting nuclear facilities and activities.

As noted, one of the most important stipulations of international norms in the area of nuclear safety is the establishment of a national legal and regulatory framework. It is clear that the two laws discussed above do not meet this requirement.[23] The Nuclear Law is not fundamentally designed to establish and allocate responsibilities for nuclear safety (incidentally, the term "safety"—*güvenlik*—does not even appear in the law). A good part of the law is about rules and procedures for the competitive selection of companies that will build nuclear power plants and price incentives to be provided for electricity produced by the plants. It does, however, stipulate the establishment of national accounts for the management of radioactive waste and decommissioning, and the requirement that nuclear power plants contribute to these funds 0.15 cents for each kWh of energy produced.

The absence of a comprehensive legal framework is partially compensated for by Turkey's adherence to a number of international agreements,[24] including the CNS. But Turkey has not yet ratified the joint convention, and this represents an important gap.

Licensing rules are laid out in the "Decree on Licensing of Nuclear Installations," dated 1983.[25] The decree defines permits and licenses that need to be obtained and authorizes TAEK to inspect facilities and enforce penalties, including the revocation of licenses. The licensing process entails three stages: site license, construction license, and operating license. There is no authorization for design. The recently published "Directive on Determination of Licensing Basis Regulations, Guides and Standards, and Reference Plant for Nuclear Power Plants, 2012," lays out a hierarchy of rules regarding licensing: it states that issues on which existing Turkish regulations do not offer sufficient clarification are to be covered "by requiring compliance of the owner/operator with the International Atomic Energy Agency (IAEA) safety documents, particularly, safety fundamentals and safety requirements."[26] Any remaining issues are to be covered by vendor-country or other third-party-country laws and regulations. This is another example of how Turkey tries to substitute international norms in areas where it has not yet developed its own regulations.

TAEK has recently issued a "Regulation on Radioactive Waste Management." It states that the management of radioactive waste is the responsibility of the person carrying out the activity that generates it. It stipulates that high-level radioactive waste shall be disposed of only in deep disposal facilities. The regulation lays out general principles for the authorization, siting, and design of radioactive waste facilities as well as their construction, commissioning, operation, and decommissioning. It does not provide for a national plan or timetables as required by the EU waste directives.

While there are still important gaps in the scope of regulations that are necessary to establish a proper regulatory framework for nuclear energy, closing the gap is technically feasible, especially given the fact that international standards and guides exist. The question is whether the overall governance structure of the regulatory framework will provide the correct incentives to the parties for the development of these regulations. Even more important is whether the correct incentives are in place for the effective implementation of the regulations. Hence it is important to review the institutional features of TAEK.

The first criterion for review is independence. Law No. 2690 gives TAEK the task of coordinating and supporting research and development activities in the field of nuclear energy. That violates the basic principle that the regulatory body should be independent from promotional activities.

There are further issues related to the independence of TAEK. According to the TAEK Law, TAEK reports to the prime minister. Yet the president of TAEK is chosen by the prime minister and appointed by a joint decision (of the prime minister, the president of the republic, and the "affiliated minister," which in this case is the minister of energy and natural resources). As was discussed, one of the internationally accepted norms of independence is that the decisionmakers in an independent regulatory authority may not be removed from duty through political discretion. To give just one example, Turkey's Law on Protection of Competition explicitly states that the president and members of the Competition Board may not be removed from duty before their term ends (Article 24). The TAEK Law does not have such a provision; the governance of TAEK is much more directly under the purview of the prime minister.

Another important dimension of independence has to do with the distribution of decisionmaking authority in regulatory affairs. Independence requires that the regulatory authority makes its decisions without undue influence. The decisionmaking structure of TAEK does not comply with this requirement. For example, the Atomic Energy Commission (AEC) has the authority to accept regulations related to TAEK. It also plays a crucial role in the licensing procedure: for example, cancellation of licenses is an AEC decision. The AEC is chaired by the president of TAEK and consists of three vice presidents (one member each from of the Ministries of National Defense, Foreign Affairs, and Energy and Natural Resources), and four faculty members engaged in educating, training, and researching in the field of nuclear energy.[27] The prime minister presides over Atomic Energy Commission meetings whenever he deems it necessary. All members are appointed by the prime minister for a four-year term, and there is no provision protecting members from discretionary removal from duty. Hence in crucial decisions, the AEC works as a subordinate of the prime minister. While this governance structure does not violate the principle of independence from bodies that "utilize or operate nuclear facilities," it does create a huge potential for undue influence, especially in a country such as Turkey where the de facto power of the prime minister can be enormous.

Another principle of independence is having adequate financial resources. The requirement here is to establish mechanisms that would limit government influence over the authority. For example, in the case of the Competition Authority, almost its entire budget is financed by special levies on the capital of newly formed companies and increases in the capital of companies. Thus the Competition Authority is completely independent in terms of financial resources. TAEK, by contrast, is part of the budget of the prime ministry and depends on yearly appropriations from the prime ministry's budgetary allocations.

The audit of the regulatory authority is also an important dimension. The preferred option here is that the agency be audited by an entity outside the direct control of the government. Many independent regulatory authorities in Turkey are audited by the Court of Accounts, which reports to the parliament. With respect to administrative and financial affairs, TAEK is under the control of the Supreme Audit Council, which is organized under the prime ministry.[28] If the need arises, TAEK may be audited by inspectors associated with the Ministry of Finance as well, upon the approval of the prime minister. So in terms of audit as well, TAEK is dependent on the political authority.

A recent change in TAEK's legal status creates further complications. As mentioned, according to Law No. 2690, TAEK was established under and reports to the prime ministry. However, a statutory decree passed in June 2011 effectively transferred the prime minister's authority over the agency to the Ministry of Energy and Natural Resources.[29] As a result, important decisions such as the appointment of the president and members of the AEC are now under the ministry's authority. The change also made some regulatory decisions and all draft regulations subject to the approval of the Ministry of Energy and Natural Resources.[30] This violates the principle that the regulatory authority should be effectively separated from other bodies or organizations concerned with the promotion or utilization of nuclear energy.

Finally, regarding transparency, it should be noted that the only statement in the TAEK Law regarding transparency consists of a vague reference to a duty to "to announce the necessary information to the public; to enlighten the public in nuclear matters."[31] Given international trends regarding transparency on regulatory matters, this, of course, is a serious omission.

It is understood that Turkey is in the process of preparing a draft Nuclear Energy and Radiation Law, which presumably is also going to establish an independent regulatory authority, or at least enhance TAEK's independence.[32] This law should have specific provisions to ensure transparency and secure the public's right of access to information. It should oblige the regulatory authority to inform the public not only of regulatory decisions but also of justifications and the reasoning that led to decisions, including background technical studies that inform them.[33] Given international developments, it should also envisage a more consultative process in licensing.

Perhaps most fundamental of all, Turkey does not yet have a comprehensive nuclear energy policy.[34] A comprehensive nuclear energy policy first of all would have presented a justification for the choice of nuclear energy, which in turn would have required an analysis of the cost and benefits of nuclear energy compared with alternative energy sources. The preparation of such a policy statement would have entailed various forms of public consultation, as well as a plan about how a comprehensive legal and operational framework for nuclear energy would be developed.

The last dimension of regulatory quality that needs to be considered concerns technical capacity and human resources. In principle this is a challenge that should be easiest to meet over the medium term, provided that it is taken seriously and adequately planned for. In the case of the planned Akkuyu nuclear plant, the Russian project company, Rosatom, has assumed responsibility for training the personnel and has brought Turkish students to Russia for education. However, Turkey needs to initiate its own plans for training and education, building on existing university programs.[35] The recent joint declaration by the prime ministers of Japan and Turkey on the establishment of a strategic partnership between the two countries reflects an intention to establish a joint international university of science and technology in Turkey that would entail, among other undertakings, training of experts in the nuclear energy field.[36]

In the short term, TAEK plans to outsource some of the critical work necessary to set up the regulatory oversight capacity for the licensing stage of the Akkuyu project. However, the technical consultancy tender has been canceled numerous times. Delays in building the necessary technical capacity may hold up the licensing of the Akkuyu plant.[37]

CONCLUSION

The regulatory framework for nuclear energy and nuclear safety in Turkey, in its present state, falls short of some internationally accepted standards and practices. Any regulatory framework for nuclear energy rests on two pillars. The first is international and supranational agreements, norms, and standards. The CNS and the joint convention are among the most important pieces of international law on nuclear safety. Turkey is a party to the first but has not yet ratified the second. While these conventions are binding, they lack enforcement mechanisms. As Turkey is a candidate country for EU accession, the EU's recent safety and waste directives are especially relevant for Turkey, and assuming that the accession process will continue, they will provide stronger incentives for as well as pressure on Turkey to build up its regulatory framework.

The second pillar of a regulatory framework is the domestic regulatory apparatus, including the existence of an independent regulatory body. The quality of the regulatory framework is closely related to the degree of independence and the extent to which mechanisms ensuring transparency and accountability are in place. The current regulatory authority, and the regulatory framework in general, does not yet satisfy the requirements of independence, transparency, and accountability. This is not only because the regulatory authority still carries out nuclear operational and developmental functions, a shortcoming that is widely acknowledged in Turkey, nor because of the recent change in TAEK's status that has placed it under the authority of the Ministry of Energy and Natural Resources. Rather, it is the whole governance structure of TAEK that jeopardizes prospects for independent decisionmaking. In terms of transparency and accountability, both TAEK and the overall regulatory framework are behind not only international benchmarks that have emerged over the past decade but are also behind other examples of independent regulatory authorities within Turkey.

Looking into the future, then, Turkey is faced with two important and highly interrelated tasks. The first is a need to complete the legislative infrastructure for nuclear energy. While this in itself is a substantial amount of work, ultimately the work is highly technical in nature and can be accomplished if sufficient resources are allocated. The second task is to set up the appropriate governance structure for regulation. This task is highly political in nature as it requires the political authority to delegate substantial power to an independent authority. At the same time, experience has shown that this second task is absolutely indispensable for nuclear safety.

One should also mention that, if anything, recent evidence seems to suggest that the current trend of the political authority is in the direction of reducing the independence of regulatory authorities, not expanding it. A 2011 law authorizes line ministries to "inspect" the activities of the agencies associated with them. While this clause does not give ministries the legal authority to overturn the decision of Turkish regulatory authorities, it does give them the ability to intimidate the regulatory authorities or exert pressure on agencies by subjecting them to inspections. At a more general level, recent developments in Turkey suggest a de facto centralization of decisionmaking authority in a very wide range in fields at the office of the prime ministry, further undermining the likelihood that the regulatory authority overseeing nuclear risks would be able to acquire or maintain sufficient actual independence.

International norms regarding the regulatory framework for safety in nuclear energy are moving in the direction of stronger independence and greater transparency. Nuclear vendors have adopted a voluntary Principles of Conduct which, among other conditions, requires signatories to make a "reasonable judgment" before entering into a contract with a country that has developed "a legislative, regulatory, and organizational infrastructure needed for implementing a safe nuclear power program" in line with IAEA standards or is in the process of developing such infrastructure.[38] Whether developments in Turkey will reflect these tendencies remains to be seen.

NOTES

1   National Diet of Japan, *The Official Report of the Fukushima Nuclear Accident Independent Investigation Commission* (Tokyo: National Diet of Japan, 2012), 16, http://reliefweb.int/sites/reliefweb.int/files/resources/NAIIC_report_lo_res2.pdf.

2   Ibid.

3   For an assessment, see İzak Atiyas, "Economic Institutions and Institutional Change in Turkey During the Neoliberal Era," *New Perspectives on Turkey* 47 (fall 2012): 45–69.

4   Selma Kuş, "International Nuclear Law in the 25 Years Between Chernobyl and Fukushima and Beyond," *Nuclear Law Bulletin* 87 (2011): 7–26.

5   For example, see Tammy De Wright, "The 'Incentive' Concept as Developed in the Nuclear Safety Conventions and Its Possible Extension to Other Sectors," *Nuclear Law Bulletin* 80 (2007): 29–47.

6   Ibid. See also Ana Stanic, "EU Law on Nuclear Safety," *Journal of Energy and Natural Resources Law* 28, no. 1 (2010): 145–58.

7   Kuş, "International Nuclear Law in the 25 Years Between Chernobyl and Fukushima and Beyond," 22.

8   International Atomic Energy Agency, *Fundamental Safety Principles, Safety Standards Series No SF-1* (Vienna: IAEA, 2006), http://www-pub.iaea.org/MTCD/publications/PDF/Pub1273_web.pdf.

9   The legal status of these fundamentals may be compared with safeguards also implemented by the IAEA. The main goal of the safeguards system is to prevent the proliferation of nuclear weapons and abuse of nuclear technology. The nuclear safeguards system is composed of a series of detailed measurements used by the IAEA for inspecting the accuracy of the reports of the countries with regard to nuclear activities and materials. Countries that have signed the safeguards agreement with the IAEA thus also accept this inspection.

10 International Atomic Energy Agency, *Milestones for the Development of a National Infrastructure for Nuclear Power*, IAEA Nuclear Energy Series, NG-G-3.1 (Vienna: IAEA, 2007), http://www-pub.iaea.org/MTCD/publications/PDF/Pub1305_web.pdf.

11 IAEA, *Governmental, Legal and Regulatory Framework for Safety*, IAEA Safety Standards Series, GSR Part 1 (Vienna: IAEA, 2010).

12 Ibid., 7–8.

13 Ibid., 30.

14 Ibid.

15 Ibid., 32.

16 Ibid.

17 International Atomic Energy Agency, *Independence in Regulatory Decision Making INSAG-17: A Report by the International Nuclear Safety Advisory Group* (Vienna: IAEA, 2003), http://www-pub.iaea.org/MTCD/publications/PDF/Pub1172_web.pdf.

18 The IAEA website describes the International Nuclear Safety Advisory Group (INSAG) as "a group of experts with high professional competence in the field of safety working in regulatory organizations, research and academic institutions and the nuclear industry. INSAG is convened under the auspices of the International Atomic Energy Agency (IAEA) with the objective to provide authoritative advice and guidance on nuclear safety approaches, policies and principles." See http://www-ns.iaea.org/committees/insag.asp.

19 IAEA, *Independence in Regulatory Decision Making INSAG-17*, 6.

20 Ibid., 9.

21 Ibid.

22 Stanic, "EU Law on Nuclear Safety," 145–58.

23 This has been emphasized in the European Commission's progress reports. For example, the 2012 progress report states: "No development can be reported on the adoption of a framework nuclear law, which would ensure a level of nuclear safety in full compliance with EU standards; existing applicable national legislation mainly covers protection against ionizing radiation and the licensing of nuclear installations." European Commission, *Commission Staff Working Document–Turkey 2012 Progress Report*, SWD, October 10, 2012, 62, http://ec.europa.eu/enlargement/pdf/key_documents/2012/package/tr_rapport_2012_en.pdf.

24 See www.taek.gov.tr/en/international/agreements.html for a list.

25 A detailed account of the various regulations regarding nuclear safety and security can be found in TAEK, "A Full Report to the 6th Review Meeting of Nuclear Safety Convention," August 2013.

26 Ibid., 3.

27 Article 6 of the TAEK Law.

28 Ibid., Article 16.

29 TAEK, "A Full Report to the 6th Review Meeting of Nuclear Safety Convention," 22.

30 Ibid.

31 TAEK Law, Article 4i.

32 TAEK, "A Full Report to the 6th Review Meeting of Nuclear Safety Convention," 22.

33 The French law on Transparency and Security in the Nuclear Field has a number of interesting provisions regarding transparency. Section 3 of the law regulates the state's duty to inform the public as well as citizens' right to access information from operators. Any disputes regarding access to information are to be resolved in administrative court. The law also stipulates specific institutional forms in the form of local information committees to ensure public access to information. See the discussion in İzak Atiyas and Deniz Sanin, "A Regulatory Authority for Nuclear Energy: Country Experiences and Proposals for Turkey," in *The Turkish Model for Transition to Nuclear Energy*, ed. Sinan Ülgen (Istanbul: EDAM, 2012).

34 See, for example, the case of the United Kingdom: Department for Business, Enterprise, and Regulatory Reform, *Meeting the Energy Challenge–A White Paper on Nuclear Power*, January 2008, www.gov.uk/government/uploads/system/uploads/attachment_data/file/228944/7296.pdf.

35 In Turkey there is one undergraduate program in nuclear energy engineering (at Hacettepe University) while several universities offer master's degrees and doctorates in the field. See Charles Ebinger, John Banks, Kevin Massy, and Govinda Avasarala, "Models for Aspirant Civil Nuclear Energy Nations in the Middle East," Energy Security Initiative at Brookings, Policy Brief 2011–1, September 2011.

36 TAEK, "A Full Report to the 6th Review Meeting of Nuclear Safety Convention," also mentions a number of projects with the IAEA and the EU planned to enhance the technical capacity of the authority.

37 Petroturk, "Akkuyu NGS'nin iş takvimi ertelenebilir" (Akkkuyu NPP Business Calendar May Be Posponed), February 7, 2014, www.petroturk.com/HaberGoster.aspx?id=10289&haber=Akkuyu-NGS-nin-is-takvimi-ertelenebilir; "Nükleer santral danışmanlık ihalesi dördüncü kez iptal" (Nuclear Power Plant Consultancy Tender Cancelled for the Fourth Time), *Zaman*, September 28, 2014, www.zaman.com.tr/ekonomi_nukleer-santral-danismanlik-ihalesi-dorduncu-kez-iptal_2143381.html.

38 See the "Nuclear Power Plant Exporters' Principles of Conduct," http://nuclearprinciples.org.

CHAPTER 3

THE ORIGINS OF TURKEY'S NUCLEAR POLICY

DORUK ERGUN

A s a country situated at the crossroads of long-lasting power struggles, military competition, and instability, Turkey has dealt with major security issues continuously throughout its existence. While there are many facets to the security challenges that Turkey faces today, one major field, and the focus of this chapter, is the role of nuclear weapons in Turkey's threat perception and as a means of response. The country's security environment has changed dramatically over the years, and this has had major impacts on how Ankara views nuclear proliferation as a threat and how it views nuclear weapons as a deterrent.

Throughout the Cold War, Turkey was almost adrift in a sea of Warsaw Pact or pro-Soviet nations and was at the heart of the nuclear competition between the United States and the Soviet Union. Although it did not develop its own nuclear weapons to deter its traditional rival, Turkey was glad to accept the North Atlantic Treaty Organization's nuclear umbrella by initially hosting Jupiter missiles for a brief period of time, and later the U.S. B61 gravity bombs. Right after their deployment, U.S. nuclear weapons gained a political meaning that Ankara has devoted much attention to by symbolizing NATO, and in particular, U.S., commitment to Turkish security and the burden-sharing principle of the alliance. Even though the bombs' strategic value and Turkey's capability to deliver them have both gradually eroded throughout the decades, their political value for Turkey has persisted.

After the demise of the Soviet Union, Turkey found itself under increasing threat from proliferation in the Middle East. Iraq, Iran, and Syria, countries that were (some still are) among top global proliferation concerns at the time, are all Turkey's neighbors, and all hold at least some level of enmity toward the country. Over the years, it became apparent that the increasingly more obsolete B61s could not satisfy Turkey's defense needs entirely. Ankara has considered moving in the direction of missile defense to complement its nuclear sharing arrangement with the United States as a deterrent against these nations but has thus far failed to take solid steps in this direction. Moreover, as a country that is an active member in international nonproliferation conventions, agreements, and institutions, Turkey has lately embraced the idea of the establishment of a Weapons of Mass Destruction–Free Zone in the Middle East—putting it in the uneasy situation of hosting NATO nuclear weapons while advocating for its adversaries to abandon theirs. Having tried to tackle the nuclear challenge over the past six decades, Ankara has increased its efforts to both facilitate diplomatic solutions and strengthen its military responses in the past five years.

Although Turkey is a longtime member of NATO, Ankara's threat perception and its views on regional politics often vary significantly from those of the rest of the allies. For one, Turkey is both a Middle Eastern country that borders Europe and a European country that borders the Middle East—making it much more susceptible to threats and developments in the region than any other NATO member. To understand Ankara's current stance and to infer its future policies thoroughly, the historical process that has led to the current state of affairs needs to be evaluated, first by inspecting Turkey's nuclear security policy during the Cold War, then by examining the proliferation challenges that Turkey has faced since the fall of the Soviet Union. Moreover, evaluating Turkey's military and diplomatic measures against more recent threats and analyzing its outstanding security challenges would be vital for anticipating the role that its nuclear policies play in the equation.

TURKISH NUCLEAR SECURITY DURING THE COLD WAR

Turkey and Russia have been engaged in an intense rivalry that can be traced back to the sixteenth century. The Czardom of Russia and later, the Russian Empire, gradually pushed the Ottoman Empire out of the Northern Black Sea and Western Caspian, supported the independence of nations in the Balkans from Ottoman rule, and scored major victories against the Ottoman Empire during World War I. This rivalry continued to be a defining factor in Turkey's threat perception and defense policies throughout the twentieth century. For example, among many other strategic, economic, and military reasons, Ankara declined to enter World War II until the very end out of fear that if Turkey joined the Axis powers, the Soviet Union would invade Turkey, and if Turkey sided with the Allies, the Soviet Union would still invade to protect or liberate Turkey from German occupation. Ankara's fears were not ungrounded. Starting from 1944, the Soviet Union did make increasingly more aggressive claims on some Turkish territories in the northeast and demanded concessions with regard to Turkey's exclusive control over the Bosphorus and Dardanelles.

Increasing Soviet assertiveness and Turkey's inability to stand on its own against Soviet pressure gradually compelled Turkey to seek the aid of the United States and join NATO. Turkey joined the alliance in 1952, at a time when the nuclear arms race between the United States and the Soviet Union was incipient. As the two superpowers raced to build more powerful nuclear explosives, and fleets of bombers, missiles, and naval platforms to deliver their ever-growing arsenals, the challenge within NATO became complicated. The United States was the primary guarantor of its allies' military security, and nuclear weapons were a principal measure of the military competition with the Soviet Union. If the Soviet Union could threaten U.S. territory with nuclear weapons, could the United States be relied upon to put its own population at risk to defend Europeans from a Soviet attack? This question—which is inherent in extended deterrence relationships—was especially worrisome for Turkey and Greece, which were not as indispensable as West Germany and other western European states, even in the eyes of most European members of NATO.[1]

To mitigate these fears and prevent European NATO members, and especially West Germany,[2] from pursuing their own nuclear programs, and to provide a credible NATO deterrent against Soviet aggression toward Europe, Washington started exploring ways to extend its nuclear umbrella to cover other NATO members. The initial plan was to form a multilateral force in which the United States would provide submarines and surface ships armed with nuclear-tipped missiles to be manned by multinational NATO crews. Turkey showed interest in hosting U.S. nuclear weapons on its territory, as did West Germany, the United Kingdom, Italy, Belgium, and Greece, and began negotiations to this end with the United States.

European allies later abandoned their support for a multilateral force when it became apparent that the United States would not give up its veto over the launch of nuclear missiles and that the financial burden of the force would be shared among all participants.[3] Instead of an alliance-wide nuclear sharing arrangement, the United States then moved to create bilateral arrangements with parties interested in hosting U.S. weapons. Following agreements with the United Kingdom and Italy, the United States and Turkey agreed in 1958 on the deployment of one Jupiter squadron in Turkey. The Turkish government agreed that the missiles would initially be manned by U.S. personnel, and the United States agreed that it would train Turkish crews to operate the missiles and would eventually relinquish control of some missiles to the Turkish authorities.[4] The Jupiters were deployed in Çiğli airbase in İzmir in 1961 and became operational a year later.

Turkey's Jupiter venture would come to be short-lived as 1962 was also the year of the Cuban Missile Crisis. In the negotiations to resolve the crisis, the United States secretly agreed to remove its missiles from Turkey in exchange for the removal of the Soviet missiles from Cuba. Turkey was never consulted throughout these negotiations.

The strategic value of the missiles was far less important than their psychological and symbolic value for Ankara. Well aware of its relative unimportance in the eyes of most NATO members, Ankara sought additional physical guarantees to underscore the commitment by NATO, and especially by the United States, to Turkish defense. In fact, even before the missiles were deployed in Turkey, the United States tried to persuade Ankara that the Jupiters were obsolete, as they took a long time to fuel and were easily detectable in the air.[5] While the United States argued that Turkey should instead depend on submarines equipped with nuclear Polaris missiles and negotiated with Ankara for the removal of the Jupiters even before the Cuban Missile Crisis, Ankara was insistent on its demand for a physical guarantee.

Another issue that might have worried Ankara was the shift in U.S. deterrence strategy that came along with the Kennedy administration in 1961, which took steps to replace the massive retaliation policy with the flexible response doctrine.[6] While in the initial policy an attack on Turkey would at least in theory mean that the United States would respond with massive nuclear force, in the flexible response doctrine, the United States called for NATO to develop an expanded and variegated range of conventional and nuclear capabilities to be able to respond more flexibly and symmetrically to possible Warsaw Pact aggression. This entailed more expenditure for conventional forces and for the deployment of large numbers of tactical nuclear weapons to the NATO theater. Use of NATO-based tactical nuclear weapons would be directed against oncoming Soviet forces, and perhaps the Soviet homeland, initially holding back the use of U.S. strategic forces. Strategists felt that this more flexible approach would provide greater credibility as a deterrent. Yet, if initiating the use of nuclear weapons with forces based in Europe (and not in the United States) was more credible (from the U.S. perspective), the strategy also made some allies question Washington's determination to fight fully on their behalf. Coupled with U.S. interest in removing the Jupiter missiles even prior to the Cuban Missile Crisis, it is probable that the Turkish authorities perceived the United States to be backing away from its commitment.[7]

Although Turkey requested alternative nuclear guarantees in exchange for the removal of the Jupiters, the United States could not provide such systems since this would run contrary to the deal reached with the Soviet Union.[8] Instead, the Kennedy administration provided Turkey with F-104G interceptors and decided to supply increasing amounts of conventional weapons.[9] Moreover, after the Jupiter missiles were removed in April 1963 with Ankara's consent, a U.S. ballistic missile submarine armed with

Polaris missiles, the USS *Sam Houston*, made a short visit to İzmir port as a show of Washington's continued commitment to Turkish security.[10]

After this period the multilateral force idea resurfaced, only to be scrapped once again since it could prove to be an obstacle against the conclusion of the Treaty on the Non-Proliferation of Nuclear Weapons (NPT).[11] The Soviet Union and the United States were cooperating in drafting and negotiating the NPT, and Moscow insisted as part of the bargain that the multilateral force concept be abandoned. Yet existing nuclear sharing arrangements within NATO were not abolished. In fact, the United States developed B61 gravity bombs in the mid-1960s and eventually deployed them at air bases in Turkey, the United Kingdom, Belgium, the Netherlands, West Germany, Greece, and Italy. The arrangement between NATO and the United States was such that the United States would train the crew of the host country to deliver the weapons and that a number of these weapons would be designated for the use of the host nation, while remaining under U.S. custody.[12]

In December 1966, a new senior body that would review NATO's nuclear policy and shape the alliance's nuclear posture was founded under the name of the Nuclear Planning Group. While membership was at first on a rotational basis, it currently includes all NATO members except for France and allows both nuclear and non-nuclear members of the alliance to take part in NATO's nuclear decisionmaking. Turkey has appreciated the Nuclear Planning Group as it allows Ankara to influence NATO's nuclear posture while building alliance cohesion and promoting burden sharing by including non-nuclear NATO members in the process.

The deployment of B61s in Europe peaked in 1971 at 7,300 gravity bombs.[13] After a few years this number began falling gradually, until reaching around 480 after the end of the Cold War. According to Hans Kristensen, in addition to hosting U.S. nuclear weapons at two national bases in Erhaç and Eskişehir, Turkey hosted NATO nuclear weapons at three bases: Akıncı and Balıkesir, each of which had the capacity to hold 24 bombs, and the İncirlik Air Base, which was able to hold 100 bombs during the Cold War. Turkey also hosted a nuclear training range in Konya.[14] After the dissolution of the Soviet Union, the number of nuclear weapons in Turkey was reduced and the weapons were consolidated at İncirlik. Balıkesir and Akıncı bases are now assumed to be in caretaker status.

Even though nuclear weapons have remained in Turkey, Ankara's worries about the solidarity of NATO and the U.S. commitment to Turkish defense have not waned. The Cyprus issue has long plagued the two countries' relationship. The late 1950s and 1960s were marked by tensions between the Turkish and Greek Cypriots of the island, the rising demand among the Greek Cypriots for uniting with Greece, the collapse of the power-sharing government of the country, and ethnic violence. As Ankara was planning to intervene in light of increased intercommunal violence in 1964, President Lyndon Johnson wrote a letter to Prime Minister Ismet İnönü that implied that the United States may not stand by Turkey if the Soviet Union struck Turkey in response to a potential Turkish intervention in Cyprus. Following a Greek ultranationalist coup d'état in 1974, Turkey invaded the northern part of Cyprus, which led to the de facto partitioning of the island. In 1975, the United States imposed an arms embargo on Turkey in response to the invasion. These U.S. actions caused Ankara to question Washington's commitment. Furthermore, during the Gulf War that followed Iraq's 1990 invasion of Kuwait, NATO was "slow and contentious"[15] in responding to Turkey's requests for air defense reinforcements to counter any potential missile attacks from Iraq on Turkish soil. NATO's hesitance raised questions in Turkey about the effectiveness and reliability of the alliance's commitment and contingency plans.

After the Soviet Union collapsed and the Warsaw Pact dissolved in the early 1990s, Turkish policymakers were briefly troubled by the prospect that NATO, and therefore the nuclear sharing arrangement, could lose its purpose. However, Ankara would soon realize that NATO's, and especially Turkey's, problems were far from over.

POST-COLD WAR SECURITY ENVIRONMENT AND PROLIFERATION THREATS SURROUNDING TURKEY

The 1991 Gulf War caught Turkish policymakers unprepared when Iraq launched retributive Scud missile strikes on Israel and Saudi Arabia. Turkey, along with the rest of the world, was already aware of the chemical arsenal and missile technologies that Iraq possessed and had used against its Kurdish minorities and against Iranian forces during the 1980–1988 war. Yet Ankara's threat perceptions were arguably based more on its adversaries' intentions than their capabilities.[16] Ankara also had a more West-centric approach in its foreign policy for most of the Cold War and paid less attention to the threats and developments in the Middle East than to the tensions and interactions between NATO and the Warsaw Pact. Yet Ankara's support for the coalition forces during the Gulf War caused fear that Turkey could suffer the same fate as Israel and Saudi Arabia and be the target of Iraqi Scud attacks. Moreover, the Gulf War revealed that the 1981 Israeli strike on Iraq's Osirak nuclear reactor did not put a stop to the Iraqi nuclear program. On the contrary, it caused Baghdad to pursue its nuclear-weapon program even more ambitiously by diversifying the locations of its facilities and employing roughly 20,000 people[17] in a program costing $7 billion to $12 billion.[18]

Iraq was not Ankara's only concern. Syria also possessed (and continues to possess) a significant stockpile of chemical weapons and had the means to deliver it. Syria and Turkey had been at odds over issues ranging from water-sharing arrangements and territorial claims to Syria's aid to the Kurdistan Workers' Party (PKK), which continues to commit terror attacks in Turkey. In fact, Ankara sent an ultimatum to Damascus in 1998 and threatened to go to war unless Syria stopped aiding the PKK and harboring its leader, Abdullah Öcalan. On October 9, Öcalan was expelled from Syria and relations between the two countries began to improve. More recently, with the Arab Spring and Ankara's open opposition to the regime of Syrian leader Bashar al-Assad, Syrian strikes using weapons of mass destruction (WMD) have once again become viable threats for Turkey.

In terms of nuclear security alone, the biggest proliferation threat emanates from Iran. Turkey and Iran have waged a rivalry for regional influence that dates back centuries. The modern rivalry between the two countries continues to play out in a variety of fields and areas, including, but not limited to, Syria, Iraq, Iran's nuclear program, Turkey's NATO membership, the Kurdish issue, and regional influence in general. In a "conventional" world, the Turco-Persian rivalry could be likened to a card game in which the sides are dealt cards from the same deck, each having similar advantages and/or disadvantages depending on the hand played. Yet Iran's *alleged* quest for nuclear weapons threatens to alter the balance entirely by bringing a completely new set of cards to the game, all of which are dealt to Tehran.

Intelligence reports suggest that Iran's clandestine nuclear program started during the shah's reign but was later brought to a halt with the 1979 revolution.[19] Yet the new regime, partly motivated by the Iran-Iraq War of 1980–1988, also started investing in nuclear technologies. According to the International Atomic Energy Agency (IAEA), Tehran made more than a dozen contacts with the A. Q. Khan network[20] between 1987 and 1999.[21] Details about Iran's nuclear program were leaked by an exiled political opposition group, the National Council of Resistance of Iran, in 2002. This forced Tehran to admit activities related to its nuclear program that it had previously hidden from the IAEA. Since then, numerous negotiations, international and unilateral sanctions, and more covert actions (possibly the assassination of Iranian nuclear scientists and cyberattacks) have failed to stop Tehran from further developing its nuclear program, which Tehran still argues is entirely peaceful. Estimates as to how long it would take Iran to achieve nuclear-weapon capability if it decides to "dash" for nuclear weapons range from a few months to over a year,[22] yet it is unclear whether Iran will choose to do so and even more dubious whether it will be able to do so before this activity is discovered by other states.

Ankara has adopted a diplomacy-first (maybe even diplomacy-only) attitude in dealing with the Iranian proliferation threat. Ankara deems economic and military measures as mostly ineffective and certainly very costly. The 1991 Gulf War and its aftermath can be taken as a precedent that has shaped Ankara's thinking. First of all, as a neighboring country, a military response creates a higher risk of retaliation compared with the coalition parties whose territories are far away from the combat zone. War also has secondary spillover effects such as major refugee influxes. Moreover, both military and economic measures hamper what is some-times quite sizable economic activity between neighbors. Ankara also argues that instead of weakening the regime, economic sanctions empower the more hardline elements in the government of the target nation, just as the economic sanctions against Saddam Hussein only helped Hussein to enrich himself and blame Western sanctions for the economic hardships endured by the Iraqi people.

Since 2008, Ankara has adopted a reconciliatory stance toward its neighbors under a policy titled "zero problems with neighbors." This included a rapprochement with Tehran. In order to gain more gravity in regional affairs and underscore its importance for both the West and the Middle East, Ankara tried to capitalize on its improving relations with Tehran and gain more clout in the negotiations between the West and Tehran by playing the role of mediator. While this policy looked promising at first, Ankara soon came under criticism from the West for its close relations with Tehran, which sometimes took the form of open advocacy of Tehran's policies. Ankara took even more heat after its relations with Israel reached a historic low in 2010 and the government started bashing Israel publicly, while at the same time making statements that were perceived as too pro-Iranian. Yet the tides that guided Turkey's relations with Iran began to turn gradually after a fuel swap deal reached by Turkey, Brazil, and Iran in 2010 was rejected by the nuclear negotiators and the UN Security Council approved a new wave of sanctions against Tehran.

Two major developments forced Turkey first to shift its involvement in the negotiations from an attempted mediator to a facilitator, and then to abandon this position altogether. The first was Turkey's decision to host NATO's early-warning radar station as part of the alliance's missile defense system. Turkey had insisted within NATO that any move to enhance missile defenses in the southern part of the alliance not be based on naming Iran as the target. Even though Ankara got its wish in the end, it failed to appease Tehran. Iran strongly resisted the NATO move and at times threatened to conduct a retaliatory strike on the early-warning radar station if Iran were to be targeted by a Western or Israeli military strike. The second issue revolved around the Syrian civil war, specifically the polar opposite policies of Iran and Turkey, which dramatically reintensified their rivalry. Even though it has been less overt since late 2013, the Turkish-Iranian rivalry is back on track, as exemplified by surfacing reports in the Turkish media of Iranian attempts to sabotage the ongoing peace process between Ankara and the PKK.[23]

Iran's development of nuclear weapons or achievement of the capability to build them would have severe impacts on Turkish foreign policy and defense. Militarily, Turkey would be forced to significantly and rapidly enhance its current WMD and ballistic missile defense capabilities and may have to gain additional conventional deterrents. In fact, Turkish policymakers have already launched an ambitious program to improve Turkey's air and missile defense capabilities.[24] Furthermore, a potential military strike against Iranian facilities would have military, political, and economic implications for Turkey. Politically, since a nuclear Iran could very likely mean that Tehran will be more assertive and bold in its foreign policy agenda, Ankara will find it harder to compete against Tehran for regional influence. Another potential impact, though neither the last nor the least, would be a regionwide renewal of competitive strategic thinking that might trigger an arms race and even compel some states to acquire their own nonconventional deterrents. Since Iran is a signatory of the NPT, any proliferation on its part could deal a heavy blow to the global non-proliferation regime and encourage other states to prepare to follow suit.

For these reasons, Turkey values the recent changes in the Iranian approach toward nuclear negotiations that came along with the 2013 election of Hassan Rouhani. Since becoming president, Rouhani has created optimism for the successful resumption of talks with the P5+1 (the five permanent members of the UN Security Council—China, France, Russia, the United Kingdom, and the United States—plus Germany) and has shown signs of a historic rapprochement with the United States. This potentially new attitude was reflected when Presidents Rouhani and Obama shared a phone call in September 2013, the first such high-level communication between Iranian and American leaders since 1979. In late November 2013, Iran and the P5+1 completed an interim agreement that basically suspended the expansion of Iran's nuclear fuel-cycle capabilities in return for partial relief of sanctions on Iran. This arrangement, which went into effect January 20, 2014, was to last for six months (with possible renewal) to allow time for the parties to negotiate a more ambitious comprehensive agreement that would operate for an extended period to be specified (between ten and twenty years, according to reports). At the time of writing, the talks have been extended twice, with a new deadline of late July 2015. For Turkey, the political and diplomatic resolution of the issue would be, without a doubt, the ideal scenario, especially considering the dreadful effects of the alternatives. The importance of success becomes ever more important with each round of talks because the passage of time potentially brings Iran closer to the bomb and hence makes military intervention even more likely. Either or both of these alternatives would have overwhelmingly negative effects for Turkey.

TURKISH RESPONSES TO THE NUCLEAR THREAT

Against this background, it appears that Turkey first started viewing the nuclear weapons and medium-range ballistic missile proliferation in the region as an actual threat in 1993. According to Ian Lesser, Ankara felt the imperative to counter proliferation in its southeast after the Gulf War: at this time, Turkey envisioned a "mix of active and passive defense against WMD," which included "reliance on NATO assets for deterrence, hardening of military targets and command and control, and bolstering the ability to locate and attack mobile targets."[25]

To bolster Turkey's defense capabilities against ballistic missiles, Turkish authorities started to explore missile defense systems. The United States did not appear to be a viable supplier because of its limited cooperation with Turkey on a number of defense issues, most importantly in the fight against the PKK, and the precedents of the U.S. Congress canceling defense contracts to Turkey on the grounds of human rights violations and its invasion of Cyprus.[26] At the same time, Turkey was reluctant to entrust its defense completely to NATO since the alliance was slow to deliver Patriot anti-missile batteries during the Gulf War, when the risk to Turkey was urgent. Ankara therefore moved to the Israeli option and began talks for the purchase of the Arrow missile defense system in 1997. The Arrow system was a preferable option for Turkey as it was specifically designed to counter Scud missiles, and particularly their Iranian variant, the Shahab missile.[27] But the Turkish-Israeli plan to co-develop a Turkish ballistic missile defense system was canceled due to the Turkish financial crisis in 2001. In 2009, Ankara announced a tender for the acquisition of a missile defense system valued at $4 billion. As will be addressed fully in chapter 5, the tender process has been delayed several times and has been the source of controversy due to Ankara's choice to lean toward a Chinese supplier for the system. Through this missile defense deal, Ankara is trying to address some of its immediate security needs and gain technological expertise and know-how to kick-start its own missile defense and ballistic missile systems.[28] As part of its domestic missile defense program, Turkey awarded a $1 billion contract in June 2011 to ASELSAN for the development of low- and medium-altitude missile defense systems.[29] Furthermore, the Turkish Air Force is on the way to establishing a Space Group Command, which will focus on satellite launches, space-based reconnaissance and imagery, and early-warning and satellite communications.[30] The country plans to invest $100 million to develop a satellite launch vehicle and plans to launch sixteen satellites by 2020 through a program valued close to $2 billion.[31] Moreover in 2011, Turkey announced plans to develop a missile with a maximum range of 2,500 kilometers—raising eyebrows about how it plans to use this technology (which is often tied to nuclear-weapon proliferation) and how it potentially can convert the technology and know-how from foreign technology transfers and satellite launch programs into attaining an offensive capability.[32]

Even if the country does not purchase or develop missile defense systems, it will still benefit from the NATO Active Layered Theater Ballistic Missile Defense (BMD) structure, which was agreed upon during the Lisbon summit in 2010. The European component of this system—the European Phased Adaptive Approach—is planned to consist of three deployment stages, in 2011, 2015, and 2018. Under this plan, an X-band early-warning radar station was deployed in Kürecik in Turkey's Malatya province and went online in January 2012. Standard Missile-3 (SM-3) interceptors are scheduled to be deployed in Romania in 2015 and in Poland in 2018. Furthermore, by 2017, the allies plan to deploy 32 Aegis BMD-capable ships, which form the backbone of the allies' BMD plans against short- and intermediate-range ballistic missiles. The entire structure will be managed from the Ramstein Air Base in Germany.[33]

Meanwhile, there are questions regarding the credibility of the deterrence provided by the presence of NATO nuclear weapons in Turkey. First, Turkey's actual capability to deliver nuclear bombs appears to have expired. During the Cold War, Turkish pilots and aircraft used to participate in exercises that allowed them to partake in nuclear strike missions. This capability seems to have eroded, as the Turkish Air Force currently takes part only in NATO's "Steadfast Moon" nuclear strike exercises and serves only an auxiliary purpose: "crews are trained in loading, unloading and employing" B61s. According to Kibaroğlu, the Turkish aircraft "serve as a non-nuclear air defense escort rather than a nuclear strike force."[34] Moreover, Turkey does not allow the U.S. Air Force to station a nuclear-capable fighter wing at İncirlik Air Base (one such request by Washington with regard to the stationing of the 52nd Fighter Wing was denied in

2005).[35] Therefore, in order for the B61s that Turkey hosts to be used in any given conflict, a U.S. fighter wing, most likely one stationed elsewhere in Europe, would have to fly to İncirlik, load the nuclear bombs, and then take off to conduct the nuclear strike. So even though Turkey possesses nuclear-capable aircraft (and will possess even more after the delivery of the 100 F-35s that Ankara has ordered from Lockheed Martin) and hosts nuclear bombs, the credibility of this deterrent has eroded considerably since the Cold War.

Therefore, speaking in terms of ballistic missile defense and nuclear deterrence, against all the potential risks in the Middle East, Turkey is currently equipped with only NATO nuclear weapons that it is not trained to use, six Patriot missile batteries that were deployed in early 2013 against the growing risks emanating from the Syrian civil war, and the NATO early-warning radar system. Thus in the event that Tehran acquires a nuclear capability, Turkey will most likely have to rely on additional NATO deterrents and defenses to satisfy its immediate defense needs. The broader political-security issue, of course, is that if Iran possessed nuclear weapons it could become more assertive in its overall discourse and actions toward Turkey and other neighbors, believing that they would be intimidated from pushing back. The asymmetry in military power potential would leave Turkey (and other neighbors) with uncomfortable options: to urge and rely on the United States and NATO to demonstrate greater resolve, backed by coercive power, to contest Iran, or to acquire greater indigenous capabilities to symmetrically or asymmetrically negate Iran's added power and thereby deter it.

While some scholars argue that Iran's proliferation could prompt Turkey to develop its own nuclear program,[36] historical trends suggest that this would be the least preferred option for Ankara and would be chosen only if Turkish security deteriorates tremendously and if all other alternatives are exhausted.[37] Traditionally speaking, Turkey has not experienced a gap in strategic parity with any of its Middle Eastern neighbors that could motivate Ankara to acquire nonconventional capabilities. Moreover, even though U.S. and NATO support has proven to be unreliable at times, Turkey has had the advantage of accessing technologically superior arms and defense technology supplied by NATO members, and hence has been able to maintain both numerical and technological conventional superiority over its

Middle Eastern neighbors. Turkey's continued defense cooperation with the United States and Israel[38] has spared Turkey from suffering a major technological gap with its neighbors, some of which briefly had access to advanced military technology after the fall of the Soviet Union. Moreover, except for the past two decades, the country has largely been uninterested in threats emanating from the Middle East, in part because its neighbors also had rivalries among themselves and all of them were weary of regional wars and conflicts. Furthermore, Turkey has arguably adopted a threat perception approach based primarily on intentions, not capabilities, and "Turkish observers find it difficult to imagine circumstances under which Iran or Iraq would employ such weapons [WMD] against Turkey—except in retaliation for American intervention launched from Turkish bases."[39] As a country so strongly entrenched in the global political framework with its NATO membership, its European Union aspirations, and its involvement in the global nonproliferation regime, Turkey would also put those vital relationships at stake if it decided to seek nuclear weapons.

ANKARA'S DIPLOMATIC RESPONSES

As will be detailed in chapter 6, Turkey is also a staunch supporter of international nonproliferation efforts. Ankara is a party to the Treaty on the Non-Proliferation of Nuclear Weapons, the Comprehensive Test Ban Treaty, the Wassenaar Arrangement, the Missile Technology Control Regime, the Zangger Committee, the Nuclear Suppliers Group, the Australia Group, the Global Initiative to Combat Nuclear Terrorism, the Seabed Arms Control Treaty, the Ottawa Convention, the Hague Code of Conduct Against Ballistic Missile Proliferation, the Proliferation Security Initiative, the Biological Weapons Convention, the Chemical Weapons Convention, and the Treaty Banning Nuclear Tests in the Atmosphere, in Outer Space and Under Water.

In addition to its active diplomatic support for international nonproliferation, Ankara in the past decade has embraced the establishment of a Middle East weapons of mass destruction–free zone. During the Cold War, Ankara viewed such nuclear-weapon–free zone efforts with reservations. Turkey regarded Soviet efforts to denuclearize the Balkans during the 1950s and 1960s as an attempt to undermine the cohesion of NATO and tip the balance of power in the its own favor[40]—moreover, any such initiative that involved the removal of U.S. nuclear weapons from Turkey would leave the country unprotected against the vastly superior Soviet army. In general, NATO members preferred to stay out of proposed WMD-free zone efforts outside of Europe, since agreeing to such initiatives could eventually invoke external and internal pressure to adhere to similar principles in their own backyard.[41]

In the past decade, and especially after 2009 when then president Abdullah Gül embraced the issue, Turkey's policies toward a WMD-free zone in the Middle East have evinced a noticeable change. While remarks from members of the current administration have at times betrayed traces of Israel bashing, Ankara's commitment to a Middle East WMD-free zone shows a genuine interest in resolving a source of long-lasting enmity and military build-up in the region. Turkey has joined the Nonproliferation and Disarmament Initiative,[42] which was formed to advance consensus outcomes of the 2010 NPT Review Conference, and called for the elimination of all nuclear weapons through this initiative. Unfortunately, international efforts to convene a conference on a Middle East WMD-free zone in 2012 did not come to fruition, as the issue is plagued with decades of mutual mistrust and enmity. Ankara does not see the effort as a challenge to its status in the NATO nuclear-weapon sharing arrangement because it sees itself as an outside actor in the Middle East.[43] This is reinforced by the fact that Turkey is not included in the geographic scope of the proposed nuclear-weapon–free zone. Moreover, the nuclear weapons stationed in Turkey are part of a wider alliance dynamic and are tied to Russian deployment of nuclear weapons in Europe, as underscored in NATO's Strategic Concept 2010.

TURKEY'S OUTSTANDING SECURITY CHALLENGES AND NUCLEAR POLICIES

In addition to the numerous proliferation challenges listed above, Turkey is faced with asymmetric and conventional threats both within and beyond its borders. Perhaps the most pressing threat at the moment is related to the ongoing Syrian civil war. In addition to the economic, social, and military challenges associated with the massive influx of refugees—the latest figure suggests that more than 1.6 million refugees are currently living in Turkey[44] with only a sixth staying in refugee camps[45]—the civil war has had numerous spillovers. Among them were: attacks on border crossings, the downing of a Turkish jet, armed conflicts with smuggler groups (some of which comprised more than a thousand individuals), the shelling of Turkish soil, and a major terrorist attack in Reyhanlı Province. Ankara has been one of the strongest opponents of the Assad regime, and this may result in further blowback from Damascus, especially if an international intervention takes place, as Ankara advocates. Failing to garner international support for an intervention so far, Ankara refrained from conducting a unilateral incursion into Syria. On two occasions—the downing of the Turkish jet and the shelling of Turkish soil—Turkey invoked Article 4 of the North Atlantic Treaty and called for consultation with the allies on these matters. While Turkey has not conducted any military operations in Syria so far, it has occasionally shelled positions of the Syrian Armed Forces. Moreover, it has provided logistical and economic support to rebel groups to hasten Assad's demise. While Turkish officials have time and again denied offering weapons to the opposition, numerous reports that have circulated in the Turkish media and elsewhere suggest that the government has at least lit a "yellow light" to the transfer of arms and weapons to rebel forces. Seeing that its strongly anti-Assad foreign policy found little support domestically and internationally, Ankara chose to tone down its opposition in the past few months. Nevertheless, it appears that the enmity between Ankara and Damascus will not cease if the Baathist regime, and especially Assad, stays in power.

Growing jihadist presence in the Syrian battlefield is another major issue for Ankara. Turkey has suffered multiple terrorist attacks from al-Qaeda, as well as other extremist organizations. The increasing presence and perhaps permanence of jihadists right across the Turkish border present many security risks for the country. The free flow of jihadists and small arms in the area presents challenges to regional stability in the long run. The foremost example of this threat is the expansion of the Islamic State in both Syria and Iraq. The instability in Syria has provided the Islamic State with ideal breeding conditions, and the organization has managed to exploit sectarian fault lines in Iraq to spread its influence and dominance. With each success, the Islamic State has managed to attract substantial numbers of foreign fighters to join its ranks and has accumulated considerable funds and military assets. The Islamic State's success on the ground stands as an example of how non-state actors may challenge the regional order and how hazardous asymmetric threats may become. As things stand at the time of writing, it appears that perfunctory measures will fall far short of stopping both the military and ideological threats that the Islamic State and other jihadist organizations present in the long run.

Meanwhile, the multiplication of Shia militias backed by Iran to counter the mounting jihadist threat presents another challenge to regional and Turkish security and stability. For Turkey, an additional major threat remains the return of the Turkish citizens fighting among the ranks of jihadist organizations, and the existing jihadist recruitment networks inside its borders, both of which may be utilized to plot attacks in the country.

Furthermore, even though Syria's declared chemical arsenal has been eliminated as agreed upon in the Kerry-Lavrov deal,[46] there remains the possibility that Syria may have residual or undeclared chemical weapons[47] or precursors that jihadists or other extremist actors could get their hands on in the near future—that is, if they have not already done so. The threat of jihadist presence thus gradually becomes a more pressing concern for Ankara—even more so than for most other members of NATO due to Turkey's proximity to the battleground. Moreover, there have already been cases in which jihadists gained control of territory along the Turkish-Syrian border, and Ankara was compelled to close border crossings due to increasing risks. In some cases, the Turkish Armed Forces shelled jihadist positions in Syria in retaliation for attacks by Islamist extremists, but Ankara has not undertaken any significant military measures to counter the impending jihadist threat so far. This situation can change, however, depending on developments on the ground.

Another trend to which Ankara has to adapt is the prospect of the foundation of an autonomous Kurdish state in northern Syria. Ankara has combated Kurdish separatism for the past three decades and has yet to accommodate Kurdish nationalism within its borders. It therefore has traditionally opposed Kurdish separatism beyond its borders in fear that this might trigger a stronger movement within Turkey. The Democratic Union Party (PYD), a Syrian-Kurdish political party, unilaterally declared its autonomy from three local governments in northern Syria in January

2014. The PYD's quest for autonomy continues to be worrisome for the Turkish leadership especially because of the PYD's close affiliation with the PKK. Ankara made some headway in tackling its own Kurdish nationalism issue by negotiating directly with PKK leader Abdullah Öcalan in a process that led to a ceasefire between Turkish Armed Forces and the PKK. Yet, at the time of writing, the process has not generated a long-lasting solution to Turkey's Kurdish issue, and the ongoing battle between the Islamic State and the PYD near Turkish borders has already exposed fault lines in the country's social fabric and the fragility of the peace process. If the peace process collapses and the PKK once again resorts to violence, lands under the PYD's control may be used as safe havens by the PKK. It is plausible to speculate that such a situation might potentially trigger a military response by Turkey.

Although Turkey was also against the establishment of a Kurdish Regional Government in northern Iraq in the aftermath of the Iraq War, Ankara has chosen to increase its economic and political relations with the Kurdish North—most of the time at the expense of the central government in Baghdad. In fact, the close alignment of Baghdad with Tehran and of Erbil with Ankara has caused some observers to view the two countries' policies in Iraq through the prism of a proxy competition with sectarian undertones. Yet regardless of the political tensions between the two sides, Iraq continues to be a source of worry and a security challenge for the region due to its instability and the sectarian violence going on within its borders. Although tensions between Ankara and Baghdad may continue in the future, it is unlikely that Iraq could pose a major military threat to Turkey in the short or medium run. Depending on the level of political tension, the sides may eventually resort to military posturing and a small-scale arms race, but full-fledged war between Turkey and Iraq seems unlikely.

As mentioned, Iran continues to present Ankara with many political and security challenges. Turkish-Iranian rivalry continues to play out in areas including Syria, Iraq, Turkey's Kurdish separatism challenge, Turkey's NATO membership, and the competition for regional influence in general. Moreover, Iran's nuclear program or an effort to thwart it may prove to be one of the biggest security issues that Turkey has faced since the founding of the republic, and the recent deal reached with Iran is still fragile. Even if a permanent agreement is reached at the negotiation table, the sides will have to overcome many other hurdles before the proliferation threat is eliminated. Even if the potential nuclear threat that Iran poses diminishes in upcoming years, it is likely that the traditional Turkish-Iranian rivalry will persist. Yet since Iran does not possess clear military superiority over Turkey, Ankara would not be compelled to seek nuclear weapons of its own in this scenario. Rather, the sides would likely choose to rely on conventional deterrents.

In addition to these acute threats, Turkey faces historical security issues that are dormant for the time being but may reemerge in the future and act as quicksand that pulls Turkey in. Some of these are: the Cyprus dispute, which has caused Turkey to resort to war and which continues to be a source of animosity between Athens and Ankara; the Nagorno-Karabakh issue and other potential sources of conflict between Azerbaijan and Armenia; and, even though relations between the sides have improved, Turkey's traditional rivalry with Russia. Moreover, as a staunch member of NATO, Turkey is automatically part of any conflict that might involve the alliance and its members in the future. The escalation of a crisis with Russia because of its annexation of Crimea demonstrates that such scenarios cannot be ruled out.

In all three contingencies regarding Syria—spillovers from the civil war, growing jihadist presence at Turkey's doorstep, and the likelihood of increasing Kurdish autonomy and influence beyond Turkey's borders—the deterrence provided by the B61 bombs and Turkey's future investments in missile defense, and, although highly unlikely, nuclear weapons of its own, would prove to be of little utility. With regard to state-level disputes with Iran, Iraq, and Syria, Ankara enjoys either military superiority or conventional parity. It would not necessarily need other means than its NATO alliance, improved conventional capabilities, and offensive and defensive missile systems to deter its neighbors. Unless Iran or any other regional rival decides to pursue nuclear weapons or poses a considerable and acute threat to Turkey with other nonconventional capabilities, Ankara would not be inclined to pursue the nuclear option.

Hence even though the security environment in and around Turkey may evolve in a way that might cause policymakers in Ankara to consider the acquisition of nuclear weapons, it is very improbable that Ankara would risk the diplomatic, political, economic, and military fallout from such an action unless it is convinced that NATO security guarantees have completely eroded and is compelled to fend for itself. The unlikely scenario of Ankara's pursuit of nuclear weapons would potentially have the adverse effect of causing other powers in the region to follow suit and proliferate, and in doing so heighten Turkey's insecurity. Therefore, while there may be plausible scenarios for Turkish proliferation in the future, Ankara does not seem to have either the need or the will to develop nuclear weapons now and is unlikely to do so in the foreseeable future.

NOTES

1   Ian O. Lesser, "Can Turkey Live With a Nuclear Iran?" On Turkey, Analysis, German Marshall Fund of the United States, March 2, 2009, www.gmfus.org/doc/Lesser_OnTurkey_0302.pdf.

2   Steven Pifer et al., "U.S. Nuclear and Extended Deterrence: Considerations and Challenges," Brookings Arms Control Series, Paper 3, Brookings Institution, May 2010, www.brookings.edu/~/media/research/files/papers/2010/6/nuclear%20deterrence/06_nuclear_deterrence.pdf.

3   Ibid.

4   "Weapons of Mass Destruction—Jupiter," Global Security, www.globalsecurity.org/wmd/systems/jupiter.htm.

5   Aaron Stein, "Turkey's NATO Nuclear Weapons History," EDAM Non-Proliferation Policy Briefs 2012/6, November 2012, http://edam.org.tr/disarmament/EN/documents/NATO%20Nuclear%20Weapons%20History.pdf.

6   Massive retaliation meant that any Soviet attack on NATO territory would be met with a massive nuclear response disproportionate to the size of the initial attack. In flexible response, the focus was shifted more onto developing additional conventional and nonconventional capabilities to counter Soviet aggression proportionately and with more flexibility.

7   Ayşegül Sever, "Yeni Bulgular Işığında 1962 Küba Krizi ve Türkiye" (Turkey and the 1962 Cuban Crisis in Light of Recent Findings), Ankara Üniversitesi SBF Journal 51, no. 1 (1997).

8   Stein, "Turkey's NATO Nuclear Weapons History."

9   Sever, "Turkey and the 1962 Cuban Crisis in Light of Recent Findings."

10 Ibid.

11 Martin Butcher et al., "NATO Nuclear Sharing and the NPT—Questions to Be Answered," Berlin Information-Center for Transatlantic Security, PENN Research Note 97.3, June 1997, www.bits.de/public/researchnote/rn97-3.htm.

12 Stein, "Turkey's NATO Nuclear Weapons History."

13 Hans M. Kristensen, "U.S. Nuclear Weapons in Europe: A Review of Post-Cold War Policy, Force Levels, and War Planning," Natural Resources Defense Council, February 2005.

14 Ibid.

15 Ian O. Lesser, "Turkey, Iran, and Nuclear Risks, " in *Getting Ready for a Nuclear-Ready Iran*, ed. Patrick Clawson and Henry Sokolski (Carlisle, Pa.: Strategic Studies Institute, October 2005), 93.

16 Lesser, "Can Turkey Live With a Nuclear Iran?"

17 Etel Solingen, "The Domestic Sources of Regional Regimes: The Evolution of Nuclear Ambiguity in the Middle East," *International Studies Quarterly* 38, no. 2 (1994): 305–37.

18 David Kay, "Iraqi Inspections: Lessons Learned," Center for Nonproliferation Studies (Winter 1993), www.acamedia.info/politics/IRef/CNS/ICrisis/lessons.htm.

19 Institute for Science and International Security, "Iran's Nuclear History From the 1950s to 2005," Institute for Science and International Security, ISIS Report, www.isisnucleariran.org/assets/pdf/Iran_Nuclear_History.pdf.

20 The A. Q. Khan network, established by top Pakistani nuclear scientist Abdul Qadeer Khan, was responsible for clandestinely transferring nuclear technology and know-how to potential proliferators including North Korea, Libya, and Iran.

21 Sharon Squassoni, "Iran's Nuclear Program: Recent Developments," CRS Report for Congress (Washington, D.C.: Congressional Research Service, 2007), http://fas.org/sgp/crs/nuke/RS21592.pdf.

22 David Albright and Christina Walrond, "Iran's Critical Capability in 2014: Verifiably Stopping Iran From Increasing the Number and Quality of Its Centrifuges," Institute for Science and International Security, July 17, 2013.

23 "İran'dan PKK'ya 'çekilmeyin' önerisi" (Iran's "Do Not Withdraw" Suggestion to PKK), NTVMSNBC, April 29, 2013, www.ntvmsnbc.com/id/25438712.

24 See chapter 5 in this volume.

25 Lesser, "Turkey, Iran, and Nuclear Risks," 94.

26 F. Stephen Larrabee, "Turkey as a U.S. Security Partner," RAND Corporation, 2008, www.rand.org/content/dam/rand/pubs/monographs/2008/RAND_MG694.pdf.

27 Aaron Stein, "Turkey Embraces Missile Defense," EDAM Non-Proliferation Policy Briefs 2012/5, November 2012.

28 Aaron Stein, "New Year's Prediction Revisited: Turkey's Missile Defense Tender," *Turkey Wonk: Nuclear and Political Musings in Turkey and Beyond* (blog), January 28, 2013, http://turkeywonk.wordpress.com/2013/01/28/new-years-prediction-revisited-turkeys-missile-defense-tender.

29 Burak E. Bekdil and Umit Enginsoy, "Aselsan Wins $1B Turkish Air Defense Contract," *Defense News*, June 23, 2011, www.defensenews.com/article/20110623/DEFSECT01/106230308/Aselsan-Wins-1B-Turkish-Air-Defense-Contract.

30 Nilsu Gören, "Turkey's Air and Missile Defense Acquisition Journey Continues," EDAM Discussion Paper Series 2013/13, Center for Economics and Foreign Policy Studies, October 2013.

31 Ibid.

32 Burak E. Bekdil, "Turkey's Sat-Launcher Plans Raise Concerns," *Defense News*, July 28, 2013, www.defensenews.com/article/20130728/DEFREG04/307280004.

33 Under the original plan, the ALTBMD consisted of four stages that would involve the deployment of improved SM-3 interceptors until 2025 and would help better tackle the intermediate-range missile threat and future ICBM threats to the United States. This stage was canceled by the United States in March 2013. For more information on the ALTBMD, please see Tom Z. Collina, "The European Phased Adaptive Approach at a Glance," Arms Control Association, May 2013, www.armscontrol.org/factsheets/Phasedadaptiveapproach.

34 Retired Turkish Air Force commander, e-mail communcation with Mustafa Kibaroğlu on April 23, 2010, quoted in Kibaroğlu, "Reassessing the Role of U.S. Nuclear Weapons in Turkey," *Arms Control Today* (June 2010): 10.

35 Federation of American Scientists, "Status of U.S. Weapons in Europe," June 26, 2008, www.fas.org/programs/ssp/nukes/_images/EuroNukes.pdf.

36 Also see chapters 7 and 8 in this volume.

37 Larrabee, "Turkey as a U.S. Security Partner."

38 Director of SIBAT (Foreign Defense Assistance and Defense Export Organization) Brig.-Gen. (res.) Shmaya Avieli has reportedly argued that Turkey's defense deals with Israel had never stopped even after the Mavi Marmara incident that caused the bilateral relations between the sides to hit an historic low. Yuval Azuali, "Arms Exports Hit Record $7.5b in 2012," Globes, July 23, 2013, www.globes.co.il/serveen/globes/docview.asp?did=1000864833&fid=1725; "İsrail: Türkiye'ye silah satışımız hiç kesilmedi" (Israel: Our Weapons Sales to Turkey Have Never Stopped), Ensonhaber.com, July 24, 2013, www.ensonhaber.com/israil-turkiyeye-silah-satisimiz-hic-kesilmedi-2013-07-24.html.

39 Lesser, "Turkey, Iran, and Nuclear Risks," 94.

40 Lykourgos Kourkouvelas, "Denuclearization on NATO's Southern Front: Allied Reactions to Soviet Proposals, 1957–1963," *Journal of Cold War Studies* 14, no. 4 (fall 2012): 197–215.

41 Ibid.

42 Current members of the organization are Australia, Canada, Chile, Germany, Japan, Mexico, the Netherlands, Poland, and the United Arab Emirates.

43 Aaron Stein, "Turkey Embraces a Middle East WMD Free Zone: Ankara's Nuclear Policies," EDAM Non-Proliferation Policy Briefs 2013/2 (January 2013), www.edam.org.tr/Media/Files/207/Policy%20Brief%20-%20TurkeyMEWMDFZAS.pdf.

44 Ceylan Yeginsu, "Turkey Strengthens Rights of Syrian Refugees," *New York Times*, December 29, 2014, www.nytimes.com/2014/12/30/world/europe/turkey-strengthens-rights-of-syrian-refugees.html.

45 "10 soruda Suriyeli mülteciler meselesi" (The Syrian Refugee Issue in 10 Questions), T24, July 30, 2014, http://t24.com.tr/haber/10-soruda-suriyeli-multeciler-meselesi,266061.

46 Thomas Gibbons-Neff, "Declared Syrian Chemical Weapons Stockpile Now Completely Destroyed," *Washington Post*, August 18, 2014, www.washingtonpost.com/news/checkpoint/wp/2014/08/18/declared-syrian-chemical-weapon-stockpile-now-completely-destroyed.

47 Amos Harel, "Israeli Intelligence: Syria Retains Small WMD Capability," *Haaretz*, October 1, 2014, www.haaretz.com/news/diplomacy-defense/.premium-1.618543.

CHAPTER 4

TURKEY'S NATIONAL SECURITY STRATEGY AND NATO NUCLEAR WEAPONS

CAN KASAPOĞLU

INTRODUCTION

Turkey's geopolitical imperatives have shifted since the collapse of the Soviet Union, from confronting a political-military giant in the north, to confronting much more diverse threats that are predominantly emanating from the Middle East.[1] In the 1990s, Ankara's security thinking was shaped by a comprehensive threat perception linked to instability in the Balkans, the Caucasus, and the Middle East. At the very beginning of that decade, Turkey's political-military elites had to face a strategic weapons threat apart from the NATO-Warsaw Pact balance of terror, namely, the ballistic missiles and chemical warheads of Saddam Hussein. The Turkish strategic community derived two lessons from this episode: first, weapons of mass destruction (WMD) threats were not only about Soviet nuclear weapons, but could also involve chemical and biological weapons in the hands of other actors. And second, although the main battle tanks and artillery of the Turkish Armed Forces look splendid during national parades, without a reliable missile defense, the nation would not be completely safe.

By the beginning of the new century, Turkish foreign policy started to shift under the influence of a new foreign policy vision that brought to the fore a neo-Ottomanist perspective. This meant giving priority to Turkey's relations with neighboring countries, especially in the Middle East.[2] During the same period, Syria and Iran had become the two main strategic weapon threats at Turkey's doorstep through their ballistic missiles as well as WMD arsenals and programs. Meanwhile, Vladimir Putin had managed to stop Russia's economic decline and had initiated a program of military modernization including the country's nuclear arsenal in order to counterbalance both the North Atlantic Treaty Organization (NATO) and a rising Beijing. Moreover, a multipolar global WMD balance emerged during the post–Cold War period with actors having the ability to operate more independently, coupled with a growing number of states possessing strategic weapons.

The shift from the Cold War threat perception to the new phases in the 1990s and the 2000s did not, however, exert a deep influence on the Turkish military strategic environment with respect to the tactical-nuclear-weapon aspect. Put simply, Turkey has continued to host NATO's tactical nuclear weapons on its soil ... but could Turkey's tactical-nuclear-weapon perspective be explained solely by NATO-Russia tactical nuclear parity?

Similar to the "nuclear taboo" of the Cold War, Turkey's security academia and strategic community were considering the deployment of tactical nuclear weapons highly symbolic without deeply questioning the B61's military value to Ankara. But if the deployment of tactical nuclear weapons on Turkish soil is highly symbolic, what could be Ankara's main guarantee against the strategic-weapon capabilities of its two problematic neighbors, Iran and Syria, especially in the absence of a viable missile defense? And in a broader context, is the "military zeitgeist" really against tactical nuclear weapons and use of low-yield nuclear weapons in conventional conflicts? This chapter aims to answer these political-military questions in order to shed light on nonstrategic or tactical nuclear weapons in the twenty-first century and Turkey's national security within the same framework.

UNDERSTANDING THE RATIONALE FOR TACTICAL NUCLEAR WEAPONS: A TURKISH PERSPECTIVE

Motivations that lead a state to acquire nuclear weapons are categorically different. Major drivers could be either security-intensive (that is, Pakistan and Israel) or "nuclear symbolism"–oriented (that is, India, North Korea, and possibly Iran). Other factors such as prestige, domestic political considerations, and regime security could be added to the list.[3] Pakistani-Indian nuclear parity is a good example of different nuclear motivations of different sides in the same conflict. Indian conventional military superiority forced Pakistani defense planners to rely on a nuclear arsenal in order to mitigate Islamabad's insecurity. At the same time, New Delhi promoted its nuclear-weapon programs in a more proactive fashion, namely, in order to foster its political-military posture in the global balance of power.[4]

At this point, the value and importance of NATO's tactical-nuclear-weapon deployment on Turkish soil and the drivers behind Ankara's decision to host B61 gravity bombs should be questioned in light of the aforementioned context. It can be stated from the outset that Ankara enjoys a clear conventional military superiority over its neighbors. The Air Force especially, as one of the major operators of the current F-16 fighter jets and the upcoming F-35 Joint Strike Fighter squadrons (a package of some 100 F-35s is planned to be acquired), along with robust armor, artillery, and mechanized units within the army coupled with a fast-modernizing navy, enables the Turkish Armed Forces to sustain a favorable military strategic balance with potential competitors. Because of this military might, Ankara's perspective on tactical nuclear weapons is expected to be different than that of Islamabad. Furthermore, unlike Israel, Turkey enjoys geostrategic depth in case of a military buildup along its borders, especially in the Second and Third Armies' areas of responsibility in eastern and southeastern Anatolia. In addition, it must be emphasized that B61 deployment on Turkish soil is subject to NATO regulations. Clearly, using these nuclear bombs in a conflict would strictly depend on a unanimous decision within NATO, so this arsenal should not be confused with a national nuclear capability. Therefore, the Turkish perspective on NATO tactical-nuclear-weapon deployment cannot be explained through any military conventional weakness–nuclear quick fix equation, as seen in many mid-sized states' experiences with strategic weapons.

Moreover, regional developments have played into the hands of Turkey in terms of military strategic balance. As a result of the U.S.-led invasion of Iraq in 2003, which reduced Saddam Hussein's military machine to an extensive internal security force, and the Syrian civil war a decade later that to a considerable extent crippled the formidable Baathist Arab army, now all that Ankara has to deal with are Turkish-Iranian and Turkish-Greek military parities[5] at its immediate borders, along with the low-intensity conflict threat emanating from the Kurdistan Workers' Party (PKK) and the recently emerging Islamic State threat along sensitive Syrian and Iraqi border areas. Apart from Turkey's territorial borders, a potential military threat could be the Russian Federation and its assertive, strategic weapons–oriented military modernization. Yet, in a broader context, the Russian strategic nuclear arsenal and widespread nonstrategic nuclear forces cannot be deterred without NATO's strategic nuclear capacity.

In addition, lessons learned from the Russian-Georgian War in 2008 and assessments of the Russian defense modernization show that tactical nuclear weapons in Moscow's arsenal are mainly planned to be used in the Eastern Military District area of responsibility in case of a military buildup and armed conflict with the Chinese People's Liberation Army, because Russian forces are outnumbered in the Far East. Indeed, Russian defense planners seem to have valid grounds when considering a nonstrategic nuclear arsenal as a military panacea to the "Chinese problem" on the Eastern front, given the fast and formidable modernization patterns of the People's Liberation Army.

This military evaluation seems to be consistent with the explanations of Turkish security experts that Ankara perceives NATO tactical-nuclear-weapon deployment as a source of prestige and a consolidator of North Atlantic security ties. However, could there be another military paradigm? Specifically, could the "over-conventionalization" in the Turkish military's strategic culture overshadow a much deeper value with the NATO tactical-nuclear-weapon deployment? The next section explores an alternative approach to Turkey's national security strategy and the role of NATO's tactical nuclear capabilities.

COULD TURKEY'S CONVENTIONAL SUPERIORITY MATCH IRAN'S AND SYRIA'S STRATEGIC WEAPONS? ASSESSING THE REAL MILITARY VALUE OF THE B61 DEPLOYMENT

The conventional wisdom is that Ankara views the forward-deployed U.S. tactical nuclear weapons on its territory as emblematic of America's commitment to its defense. But perhaps another, as yet underexplored, reasoning is behind Ankara's commitment to continue to host these weapons: insecurity emanating from Iran and/or Syria.

Indeed, although Ankara enjoys conventional superiority over its existing and potential regional challengers, Turkey's security environment, especially in the Middle East, has been shifting away from the conventional military paradigm. Military trends suggest that the future threat landscape in the Middle East and North Africa region will in all likelihood be dominated by asymmetric conflicts, as well as strategic weapon systems and their intrawar deterrence functions. Within this context, Turkey is losing not its capability to defend against territorial incursions, but rather its capacity of intrawar deterrence.

W. Andrew Terrill of the U.S. Army's Strategic Studies Institute explains the concept of intrawar deterrence as "the effort to control substantial military escalation during an ongoing war through the threat of large-scale and usually nuclear retaliation should the adversary escalate a conflict beyond a particularly important threshold."[6] Starting with the 1973 Yom Kippur War, which saw the first Scud short-range ballistic missile launch by the Egyptians along with the reported Israeli preparation for the nuclear option in case of an existential threat,[7] up until the more recent examples of tactical chemical weapon use in the Iran-Iraq War and the Syrian civil war, the Middle Eastern battleground proved the functionality and military value of strategic weapon systems. Furthermore, in the Middle East, acquiring strategic weapons is seen as a "quick-fix" solution for conventional military shortcomings, as well as a regime guarantee.[8]

In this context, Ankara needs to adapt to the new regional military strategic parameters in order to sustain its regional assertions. Put simply, the Syrian civil war showed that despite Turkey's conventional superiority, the combination of ballistic missiles and chemical weapons at the hands of Syrian leader Bashar al-Assad has restricted Ankara's intervention capabilities in a regional contingency. What is more, although the Syrian regime has been facing one of the most destructive civil war conditions, it managed, through a carefully tailored escalation strategy, to use its chemical arsenal. It can be argued that Assad's use of chemical weapons deterred both the Turkish unilateral option and the allied military option. Thus, the Turkish military posture, and the role of tactical nuclear weapons, should be seen through the prism of current military trends and future security needs, and a holistic analytical approach to Turkish defense modernization must be adopted.

TURKISH MILITARY MODERNIZATION: KEEPING CONVENTIONAL SUPERIORITY AND SEARCHING FOR STRATEGIC DEFENSIVE CAPABILITIES

By the 2000s, some landmark projects of the Turkish defense industries and procurement programs have come into prominence. For instance, the multibillion-dollar long-range air and missile defense systems (T-Loramids) project, and Ankara's controversial attempts to procure the Chinese HQ-9 system, are among the most important issues on the Turkish defense agenda at the time of writing. The TF-2000 Project, which aims to procure anti-air warfare frigates for the Turkish Navy, is designed to foster the Turkish Armed Forces' naval-based air and missile defense and electronic warfare capabilities.[9] In combination with the T-Loramids, the TF-2000 Project is expected to improve Turkey's strategic defensive capabilities and enhance the mobility of its strategic defenses by taking advantage of the nation's long coast lines between the Black Sea and the Eastern Mediterranean. Furthermore, in case Ankara manages to successfully integrate land-, naval-, and air-based systems, in conjunction with the NATO capabilities in this field, then tracking incoming threats and cueing information between different layers of strategic defense would provide a significant defensive shield for Turkey. For starters, we have seen a larger-scale and more sophisticated model of this military ratio in the integration of Aegis ballistic missile defense system–equipped destroyers and cruisers with land-based systems (that is, Patriot missiles and terminal high altitude area defense, or THAAD), as well as space-based ISTAR (intelligence, surveillance, target acquisition, and reconnaissance) assets.[10]

Along with these aspects, Ankara is working on enhancing its offensive counter-air and air-ground capabilities. The F-35 Joint Strike Fighter project, at which Turkey is a Level 3 participant in the consortium, is expected to replace Turkey's aging F-4 and F-16 squadrons. The multi-role fighter will increase the Turkish Air Force's offensive and standoff strike capacity through its air-ground and stealth features. In this sense, the Turkish defense industry has been working for a few years on developing a national cruise missile, the SOM-J, with a vision to integrate the standoff weapon system with the F-35s on order. Turkey's SOM missile is able to carry a 200 kilogram warhead within a 50–200 kilometer range, and it is subject to further modernization.[11] In this regard, at the time of writing, Lockheed Martin and Roketsan signed an agreement to develop a 1,000 pound class derivative of the initial SOM missile for Turkey's F-35 Block-IV.[12]

All of these projects give an important hint about the underlying military rationale behind Ankara's defense modernization plans. Clearly, Ankara's efforts have been focusing on developing robust defensive strategic weapons capabilities in order to minimize its strategic vulnerability against offensive strategic weapons. In other words, Turkey has already been involved in the strategic arms race in the Middle East in one way or another.

In addition to Ankara's efforts at improving its defensive strategic capabilities and integrated air and missile defense capacity, the army keeps progressing with a conventional military paradigm. The army, or the Turkish Land Forces as it is known in the Turkish doctrinal order of battle, constitutes the backbone of the Turkish Armed Forces. It has been under restructuring and modernization under the Kuvvet 14 (Force 14) program in the recent years. The Kuvvet 14 is designed to reduce the army by 20–30 percent in terms of personnel and unit size but to improve its capabilities in rapid deployment, mobility, and firepower. Within the program, Turkey is building its land-aviation capacity with the joint production of T129 attack helicopters by Turkish Aerospace Industries and AgustaWestland.[13] In addition, since the 1990s, the Turkish Land Forces' doctrinal order of battle has been to replace a division-based system with a more mobile and flexible corps-brigade-battalion structure. Furthermore, Turkish defense modernization efforts have included ambitious national main battle tank, national corvette, and national drone projects. However, although these developments reflect a significant military uptrend in Turkish defense modernization, offensive strategic weapons capabilities are still lacking.

REGIONAL CONTENDERS: IRAN AND SYRIA

When it comes to Iranian military capabilities, we see both significant weaknesses and strengths at the same time. According to a 2008 U.S. Congressional Research Service report, Iran's military would be effective against its weaker neighbors (postwar Iraq, Turkmenistan, Azerbaijan, and Afghanistan) but less capable against Turkey and Pakistan.[14] But one of the most critical aspects of the Iranian military modernization is strategic weapon systems, especially ballistic missile proliferation.

Tehran is working on new ballistic missile systems in order to improve the effectiveness of its Shahab class ballistic missiles. Iran also owns Ghadir-1 and Sajjil-2 class missiles. These are classified as medium-range ballistic missiles that can strike targets from distances between 1,000 and 3,000 kilometers. Moreover, both weapon systems are able to deliver conventional warheads with a payload of about 1,000 kilograms, and, theoretically, nuclear warheads and chemical submunitions as well. The Ghadir-1, an improved version of the Shahab-3, is reported to reduce the Shahab line's problematic circular error probability from 2,500 meters to some 1,000 meters. Furthermore, the improved triconic-shaped warhead of the Ghadir-1 is believed to increase the accuracy and stability of the reentry vehicle with a higher payload.[15]

The recently developed Sajjil-2 is a solid-propellant missile that can target distances of 2,000–2,500 kilometers and as such poses a significant threat to Turkey and the region. Solid-propellant missiles, which are maintained at high readiness levels, decrease the launch cycle, raising the possibility of a surprise attack. Furthermore, the Sajjil-2 is a swiftly deployable system that can be carried by road-mobile transporter erector launchers. This capability allows Tehran to target all six Gulf Cooperation Council states (Bahrain, Kuwait, Oman, Qatar, Saudi Arabia, and the United Arab Emirates), plus the entire territories of Turkey, Israel, Jordan, Lebanon, Syria, and Iraq, as well as the island of Cyprus where Turkey stations a corps-level deployment. Combined with its road-mobile character and fewer personnel needs, the solid-fuel Sajjil-2 arsenal could prove to be pretty resilient against first and preventive strikes in a possible conflict.

But the most important point in the Iranian defense modernization as it relates to the Turkish-Iranian military balance is the cumulative know-how that is necessary to produce such a strategic weapon system. Technically, designing and manufacturing solid propellant engines and fuels for missiles require more sophisticated efforts than for liquid propellant missiles. As such, the Iranian missile proliferation trend and increasing know-how should be taken as a more pressing threat than their actual capabilities for now, as they show the future trajectory of the Turkish-Iranian military balance. Finally, it is not only the military dimension of Iran's nuclear program, but also the still unknown scope of its WMD research that could pose a significant threat to Turkey. For instance, although Iran is a party to the Biological Weapons Convention, it has failed to comply with its obligation to declare national biological defense research programs as well as its past activities in offensive and defense biological research and development programs. Furthermore, some statements by top Iranian figures and Tehran's dual-use pharmaceutical activities create concerns about its military-use biotech capacity.[16] Similarly, although Tehran is a member of the Chemical Weapons Convention, there are critical reports about its chemical weapons research and development. For instance, the Nuclear Threat Initiative indicates that

> Iran has a sophisticated base for the development of a chemical warfare (CW) program that dates back to the Iran-Iraq War (1980 to 1988)—a conflict that gave Iran strong incentives for developing a robust chemical defense capability. Several unclassified allegations and reports from the 1990s until approximately 2003 suggested that Iran had developed an offensive CW program. Many of these claims described specific military capabilities related to agent stockpiles, delivery systems, and deployments that cannot be independently verified in open sources.[17]

From Turkey's point of view, the biggest caveat is the uncertainty about Tehran's strategic weapons and know-how that shapes Ankara's security thinking about Iran and undermines its sense of military superiority. From a military perspective, tactical nuclear weapons could be seen as an instrument for keeping military parity with a rival that possesses non-nuclear strategic weapons such as chemical or biological assets.[18] Therefore, a good reason to explain Ankara's desire to keep NATO's tactical-nuclear-weapon assets on its soil could be the Iranian strategic weapons capabilities.

In this regard, Syria also poses a potential threat to Turkey. Even after the deal on the removal of the regime's chemical weapons, and assuming that Assad has been running a fair and transparent disarmament campaign and would not resume Syria's chemical weapons program in the near future, the Syrian regime will continue to own a combination of a notorious biological weapons program and ballistic missiles. Anthrax and cholera bacteria, along with botulinum and ricin toxins, are said to constitute the backbone of the Syrian biological weapons arsenal.[19] Weaponization of biological agents is not easy. However, a Center for Strategic and International Studies (CSIS) report indicates that the technological knowhow, which would be adequate for producing bomblets and cluster munitions for the VX chemical agent, which Syria has already managed to weaponize, can be adapted to weaponize biological agents. In this regard, the CSIS report notes that dry micropowdered advanced bio-agents, such as the lethal forms of anthrax, would be similar to small theater nuclear bombs in terms of destructiveness.[20] Besides, anthrax bacteria spores are very resistant to extreme temperature conditions and can also be used to contaminate water and food. With an incubation period from hours to seven days, weaponized anthrax's effects could be devastating in large urban areas.[21] Therefore, in case of a considerable chemical disarmament of Syria, the residual biological weapons arsenal of Damascus would still pose a threat to Ankara. Yet, the only nonconventional response that could counterbalance this threat is NATO's tactical nuclear-weapon deployment on Turkish soil.

A military capability such as tactical nuclear weapons would not only alter the combat calculus in a war situation, but would also provide an intrawar deterrence to control military escalations and political crises. Moreover, strategic weapons are capable of creating strategic consequences without being actually employed. Therefore a combination of the Washington Treaty's Article 5 guarantees, which ensure collective defense for NATO members, and tangible deterrence provided by the forward-deployed B61 assets could help to deter possible aggression.

GETTING THE CONTEXT RIGHT: BETWEEN THE COLD WAR NIGHTMARES AND TODAY'S DREAMS

Despite the alternative perspective on the possible roles of NATO's tactical nuclear weapons in Turkey, their deployment itself was primarily initiated against Moscow's expansionist threat during the Cold War. As of February 2011, when the New START Treaty entered into force, it was estimated that some 150–200 tactical nuclear weapons have been deployed in the territory of five European NATO allies: Germany, Italy, the Netherlands, Belgium, and Turkey. According to several open-source publications, Turkey is believed to host some 60–70 of those assets at the İncirlik Base.[22] On the other side, Russia is believed to still retain about 2,000 operational tactical nuclear weapons with other yet nonoperational assets. Unlike NATO's tactical-nuclear-weapon posture in Europe, Moscow enjoys an operational diversity in delivery means via aircraft that are capable of carrying air-to-surface missiles and gravity bombs (that is, the TU-22 and SU-24), surface-to-air missiles (the SA-10), submarine-launched assets (SS-N-9, SS-N-12), and ballistic missiles (SS-26 Iskander). More important, the Russian military considers tactical nuclear weapons as a means of compensating for Moscow's conventional handicaps in Europe and a response to NATO's increasing ballistic missile defense capabilities and the alliance's further enlargement, as well as a reliable asset in a military buildup scenario against China.[23]

The tactical-nuclear-weapon presence in Turkey, though, is not a response to the recent doctrinal shifts in Russia but a true Cold War strategic legacy. By the mid-1980s, Washington had deployed some 500 warheads in Turkey, scattered at four air bases. At that time, Turkish military posture was able and ready to conduct nuclear missions using F-104, F-4, and F-100 fixed-wing assets, as well as via some (about four battalions) nuclear-capable land force units. Since the end of the Cold War, Turkey continues to host tactical nuclear capability in smaller numbers. Yet, Turkish officials and experts claim that the Turkish Air Force no longer conducts military exercises with B61 type (and variant) tactical nuclear weapons.[24] According to this perspective, as mentioned, a majority of Turkish security experts claim that Ankara still, decades later, considers the deployment an "emblem" of prestige within NATO and also a strong tie to the Western security umbrella.[25] Some foreign experts, meanwhile, opine that the Turkish Air Force's nuclear mission capabilities could still be more than a Cold War story. According to Hans Kristensen,

> U.S. presidential deployment authorization for Europe issued in December 2001 still included 40 "host" weapons for delivery by the Turkish Air Force [and] approximately 30 F-16C/D Block 50s are scheduled to receive a "stop-gap" upgrade to make them capable of carrying the new B61-12 bomb that will replace the B61-3/4 beginning in 2019.[26]

At this point, Turkey's F-35 Joint Strike Fighter acquisition should be assessed carefully, as its possible dual-capable aspect could bring about more than an offensive counter-air and air-ground role to the Turkish Air Force's capabilities with respect to tactical nuclear weapons. Turkey is a Level 3 participant of the Joint Strike Fighter project and would acquire about 100 conventional takeoff and landing F-35As to its air wing.[27] The first delivery of the acquisition is scheduled to take place in 2018.[28] Some experts underline the very fact that in the wake of the U.S. Department of Defense's 2010 Nuclear Posture Review, the future of tactical nuclear weapons depends on the combination of the success of the F-35 project and the refurbishment of the B61.[29] Similarly, a 2013 RAND Corporation report claims that the F-35 platform and B61 tactical nuclear weapons will constitute the tactical delivery leg of the U.S. Air Force's future nuclear missions.[30] Although it is unknown at this stage whether the Turkish F-35 tactical wing would be nuclear certified or nuclear mission ready, a combination of F-35 squadrons and B61 presence could provide a wide array of options and flexible deterrence to Ankara.

Meanwhile, NATO's military thinking is shifting from favoring tactical-nuclear-weapon deployment in Europe, and this could bring about significant consequences to Turkey's national security paradigm. For instance, unlike NATO's 1991 and 1999 Strategic Concept documents, the latest

Strategic Concept, adopted at a summit in Lisbon in 2010, did not mention "sub-strategic nuclear weapons in Europe."[31] In fact, the 1991 document openly emphasized the role of sub-strategic nuclear assets in Europe as a link to the alliance's strategic nuclear forces and also as a reinforcing element that augments transatlantic ties. The second publicly open Strategic Concept of the alliance, which was released in 1999, reiterated the same expression.[32] Thus, following the 2010 Strategic Concept, "a new link to the alliance's strategic nuclear forces," and means of reinforcing transatlantic ties might be needed in the near future. Clearly, although the 2010 Strategic Concept indicated that NATO will remain nuclear as long as nuclear weapons exist, the Lisbon document does not mention sub-strategic nuclear assets, and underlines that

> The supreme guarantee of the security of the Allies is provided by the strategic nuclear forces of the Alliance, particularly those of the United States; the independent strategic nuclear forces of the United Kingdom and France, which have a deterrent role of their own, contribute to the overall deterrence and security of the Allies.[33]

Yet, the U.S. Nuclear Posture Review draws a more positive stance toward tactical nuclear weapons by indicating that the United States will "retain the capability to forward-deploy U.S. nuclear weapons on tactical fighter-bombers and heavy bombers, and proceed with full scope life extension for the B61 bomb including enhancing safety, security, and use control."[34]

Still, it may be too early to state unequivocally that the era of NATO tactical nuclear posture is over. Technically, a military capability could be abandoned if the perceived threat is altered or eliminated, or if a new system takes the place of the existing one. Considering the doctrinal shifts and modernization trends in the Russian Federation, as well as other NATO strategic weapon–capable challengers, it cannot be ascertained that the threat is gone. Yet, in recent years, the strategic community has debated whether a new system, NATO missile defense, could be compensation for the possible removal of the tactical nuclear weapons from Europe.

COULD MISSILE DEFENSE COMPENSATE FOR NATO'S TACTICAL NUCLEAR WEAPONS? UNDERSTANDING TURKEY'S KEY BUT AMBIGUOUS POSITION

At the 2010 Lisbon summit, NATO leaders approved the European-based territorial ballistic missile defense plan that would be integrated with U.S. ballistic missile defense capabilities. At the 2012 Chicago summit, NATO declared that it had reached an "interim capability." According to some analysts, at a time when tactical-nuclear-weapon deployment in Europe comes under question, NATO's missile defense could be a compensation for withdrawal of tactical nuclear weapons from Europe as a means of reassuring U.S. commitment to the alliance's defense.[35] From a military standpoint, the main issue appears to be whether tactical nuclear weapons and a missile defense shield should be substitute or complementary assets and what Turkey's position could be in this regard.

The increasing need for ballistic missile defense capabilities is not ema- nating solely from ballistic missiles, but from a combination of ballistic missiles and WMD threats and from advantages in military technology and defense economics. First, ballistic missiles enjoy fewer maintenance, training, and logistics requirements than manned aircraft. And second, these weapon systems can be armed with chemical, biological, and nuclear warheads. In terms of defense economics and politico-technical restraints, countries that cannot sustain nuclear programs usually opt for chemical and biological warheads that are cheaper and easier to obtain. Furthermore, biological and chemical submunitions in ballistic missile warheads make accuracy problems less important against urban areas and large military concentrations due to their wide dispersion capabilities.[36] NATO officially draws attention to ballistic missile and WMD warhead proliferation together as the grounds for the allied missile defense efforts.[37]

With regard to debates over tactical nuclear weapons versus ballistic missile defense for NATO's future force posture, Turkey's position is criti- cal and of interest. Turkey is an important NATO member that hosts tactical nuclear weapons on its soil under the burden-sharing concept, and it is a key component of the missile defense shield, with the X-band early warning radar having been installed in Kürecik in Turkey's Malatya dis- trict. Turkey is also the sole NATO nation bordering Iran, one of the unnamed targets of the missile defense shield. Finally, Ankara is in the midst of its own ballistic missile defense procurement plan, the T-Loramids project, in which Turkish decisionmakers were initially edging toward a decision to buy a Chinese-made system despite competing bids by Turkey's NATO allies including the U.S. Patriot and the European Aster 30. Nevertheless, as the tender was extended a few times and given the pressure on Ankara by NATO allies, it would be fair to say that the Patriot and Aster 30 options now could have a better chance.

Two critical issues loom large with regard to Turkey's position in NATO missile defense and the importance of the X-band radar in Kürecik. The first issue is data networking and integration. Clearly, in the absence of full integration of the X-band radar in Turkey and planned interceptors in Poland and Romania, it would not be possible to run the missile defense system efficiently.[38] The second issue is the protection of the X-band radar in case of a conflict situation, especially against potential missile strikes. Such protection would necessitate point missile defense, such as Patriot batteries.[39] Notably, at the time of writing, Ankara has been working on a procurement deal with Beijing for the HQ-9 system. If Turkey ends up with this procurement project despite objections from its NATO allies, including Washington, the Chinese ballistic missile defense system—a non-NATO asset—will be expected to provide protection for one of the most critical parts of NATO's biggest defense initiative in the twenty-first century.

Although NATO missile defense is seen by some as a replacement for the tactical-nuclear-weapon deployment in Europe, the essence of ballistic missile defense systems and tactical nuclear weapons is different from a military standpoint. Thus, they would be best used in harmony rather than compensation. Although an effective ballistic missile defense system would significantly minimize strategic vulnerability and deny offensive advantages to states potentially holding WMD or ballistic missiles, the nature of ballistic missile defense has very little capability in retaliatory action, coercive diplomacy endeavors, preemptive military missions, and projecting active deterrence. At the same time, when tactical air wings are used as tactical-nuclear-weapon delivery platforms—which is the existing NATO concept—having bases close to where the weapons would be delivered constitutes a main order of battle. This close basing provides higher sortie rates and enables surprise attacks, but it also is vulnerable to adversaries that possess formidable ballistic missile arsenals. When applied to the Turkish tactical-nuclear-weapon case, given Iranian and Syrian offensive strategic weapons capabilities, the tactical-nuclear-weapon and possible dual-capable aircraft deployment in Turkey would strongly require ballistic missile defense protection. In return, these offensive strategic capabilities could be matched and deterred only by either high-end precision strike conventional assets or nonconventional offensive capabilities. Therefore, Turkish national defense would need a combination of integrated NATO and national ballistic missile defense systems, preferably an offer by one of the NATO members in the T-Loramids tender, and tactical-nuclear-weapon deployment at the same time. Clearly Ankara would need both offensive and defensive strategic capabilities that go well beyond its conventional superiorities as long as:

- Turkey remains under the missile range of its competitors in the Middle East

- The current regional missile trends continue in terms of solid-fuel systems, which minimize the launch cycle (the increase in quantity would enable ballistic missile defense–saturating launch salvos, and the increase in range would foster the production of medium-range ballistic missiles)

- Turkey's regional rivals pursue their existing (or suspected) WMD arsenals and reported WMD research and development programs

- A comprehensive arms control regime is absent in the Middle East, to include full implementation of the Treaty on the Non-Proliferation of Nuclear Weapons, Chemical Weapons Convention, Biological Weapons Convention, and Missile Technology Control Regime

CONCLUSION

Turkey's nuclear stance within NATO has generally been seen through the prism of national prestige and the consolidation of North Atlantic security ties. But a deeper insight into the Turkish case and Ankara's delicate military strategic balance in its hinterland with respect to strategic weapon systems uncovers a more complex rationale beyond the generally accepted motive of commitment to NATO's burden sharing and nuclear posture. The positioning of NATO's tactical nuclear capabilities on Turkish soil has more than a symbolic meaning for Ankara. Turkey's nuclear stance within NATO is linked to threat perceptions emanating from strategic weapons proliferation in the Middle East.

This observation leads to a number of conclusions regarding Turkey's stance with respect to the future of nuclear weapons within NATO.

Even under the scenario of a possible inclination for tactical-nuclear-weapon disarmament, Turkey's stance should be expected to be different from that of its European NATO allies. First, Ankara has to deal with ballistic missile and WMD proliferation by rogue states at its doorstep. In particular, chemical warfare carried out in Middle Eastern conflicts (the Iran-Iraq War, the genocidal al-Anfal Campaign, the Syrian civil war) and notorious biological weapons programs, which have loomed large with the Syrian case and Saddam Hussein's military-purpose research in Iraq, have been posing a tangible and menacing threat. Second, due to the Syrian civil war, the region's security environment has led to a sectarian polarization threat landscape in which Ankara finds itself in a fierce power struggle with the two strategic weapons holders, namely the Baathist regime of Syria and Iran backstage. Third, given the ballistic missile proliferation and military trends toward acquiring strategic weapons as quick-fix solutions in the Middle East, Turkey's geopolitical imperative is minimizing its strategic vulnerabilities vis-à-vis its existing and potential competitors. Especially considering Ankara's recent assertive policies in the region and its push for securing a leadership role, any Turkish administration would be under the constant challenge of a threat emanating from WMDs and ballistic missiles.

Neither the further development of NATO's ballistic missile defense system nor the improvement of Turkey's own capabilities relating to ballistic missile defense is likely to change the calculus of the authorities in Ankara.

To avoid fostering already problematic WMD proliferation at their Middle Eastern doorstep, Turkish decisionmakers are not able to voice their open and powerful support for the NATO tactical-nuclear-weapon presence in Turkey and continuation of the burden-sharing concept. Yet, Ankara's national security imperatives necessitate a robust intrawar deterrence that could be maintained solely by such weapon systems that can counterbalance offensive strategic weapon capabilities at the hands of Ankara's regional competitors. In this regard, NATO's tactical-nuclear-weapon assets clearly mean more than nuclear symbolism and allied commitments to Turkey's national security.

NOTES

1   F. Stephen Larrabee and Alireza Nader, *Turkish-Iranian Relations in a Changing Middle East* (Santa Monica, Calif.: RAND Corporation, 2013), 5.

2   Omer Taspinar, *Turkey's Middle East Policies: Between Neo-Ottomanism and Kemalism*, Carnegie Paper, Carnegie Endowment for International Peace, 2008, 15.

3   Adil Sultan, "Pakistan's Emerging Nuclear Posture: Impact of Drivers and Technology on Nuclear Doctrine," *Strategic Studies Journal* 31–32 (Winter 2011–Spring 2012): 147–67.

4   Ibid., 150.

5   As Athens is a member of the North Atlantic Alliance, NATO tactical-nuclear-weapon deployment on Turkish soil cannot be counted as a military asset in the Turkish-Greek military balance.

6   W. Andrew Terrill, *Escalation and Intrawar Deterrence During Limited Wars in the Middle East* (Carlisle, Pa.: Strategic Studies Institute, 2009), xi.

7   Ibid., 44.

8   Richard L. Russel, "The Middle East's Nuclear Future," in *The Next Arms Race*, ed. Henry D. Sokolski (Carlisle, Pa.: Strategic Studies Institute, 2012), 186.

9   For detailed information visit the official webpage of the under secretary for defense industries, www.ssm.gov.tr.

10 "Aegis Ballistic Missile Defense Fact Sheet," Lockheed Martin, 2013.

11 For detailed information visit http://missilethreat.com/missiles/stand-off-missile-som.

12 Marina Malenic, "Lockheed Martin Announces Roketsan Teaming on New F-35 Standoff Missile," *IHS Jane's Defence Weekly*, October 22, 2014, www.janes.com/article/44921/lockheed-martin-announces-roketsan-teaming-on-new-f-35-standoff-missile.

13 IHS Jane's, "World's Armies–Turkey," 2012, 9.

14 IHS Jane's, *Jane's Sentinel Security Assessment—The Gulf States: Iran Armed Forces*, 2012, 1.

15 Anthony Cordesman and Bryan Gold, *The Gulf Military Balance Volume II: The Missile and Nuclear Dimensions* (Washington, D.C.: Center for Strategic and International Studies, 2013), 55–56.

16 Anthony Cordesman and Adam Seitz, *Iranian Weapons of Mass Destruction: Biological Weapons Programs* (Washington, D.C.: Center for Strategic and International Studies, 2008), 7.

17 Nuclear Threat Initiative, "Country Profiles: Iran-Chemical," www.nti.org/country-profiles/iran.

18 Hugh Beach, *Tactical Nuclear Weapons: Europe's Redundant Weapons of Mass Destruction* (London: International Security Information Service, 2004), 14.

19 Dany Shoham. "The Fate of Syria's Chemical and Biological Weapons," BESA Center Perspective Paper 177, Besa Center, 2012, http://besacenter.org/perspectives-papers/the-fate-of-syrias-chemical-and-biological-weapons.

20 Anthony Cordesman, *Syrian Weapons of Mass Destruction* (Washington, D.C.: Center for Strategic and International Studies, 2008), 15.

21 For detailed information see Elizabeth M. Whelan et al., "Anthrax: What You Need to Know," fact sheet, American Council on Science and Health, 2003.

22 Götz Neuneck, "European and German Perspectives," in *Tactical Nuclear Weapons and NATO*, ed. Tom Nichols, Douglas Stuart, and Jeffrey D. McCausland (Carlisle, Pa.: Strategic Studies Institute, 2012), 259–63.

23 Ibid., 262–65.

24 Mustafa Kibaroğlu. "Turkey NATO and Nuclear Sharing: Prospects After NATO's Lisbon Summit," Nuclear Policy Paper no. 5 (Reducing the Role of Tactical Weapons in Europe Project), British-American Security Information Council, 2011, 2–3, www.basicint.org/sites/default/files/Nuclear_Policy_Paper_No5.pdf.

25 Ibid.

26 Hans Kristensen, *Non-Strategic Nuclear Weapons* (Washington, D.C.: Federation of American Scientists, 2012), 21–22.

27 IHS Jane's, "World's Air Forces–Turkey," 2012, 3.

28 Lale Sariibrahimoglu, "Turkey Recommits to F-35 Programme," *IHS Jane's Defence Weekly*, May 7, 2014, www.janes.com/article/37547/turkey-recommits-to-f-35-programme.

29 Rebecca Grant, "Nukes for NATO," *Air Force Magazine* (2010): 43.

30 Don Snyder et al., *Sustaining the U.S. Air Force Nuclear Mission* (Santa Monica, Calif.: RAND Corporation, 2013), 7.

31 North Atlantic Treaty Organization, *Strategic Concept: Active Engagement, Modern Defence*, 2010.

32 North Atlantic Treaty Organization, *Strategic Concept*, 1991, Article 56; *Strategic Concept*, 1999, Article 64.

33 North Atlantic Treaty Organization, *Strategic Concept: Active Engagement, Modern Defence*, 2010.

34 U.S. Department of Defense, *Nuclear Posture Review 2010*, xiii.

35 Steven A. Hildreth and Carl Ek, *Missile Defense and NATO's Lisbon Summit* (Washington, D.C.: Congressional Research Service, 2011), 7.

36 National Air and Space Intelligence Center, *Ballistic and Cruise Missile Threat*, NASIC, 2013, 4.

37 NATO Public Diplomacy Division, "Missile Defense Fact Sheet," June 21, 2011.

38 Steven J. Whitmore and John R. Deni, *European Phased Adaptive Approach: The Implications of Burden-Sharing and the Underappreciated Role of the U.S. Army* (Carlisle, Pa.: Strategic Studies Institute, 2013), 16.

39 Ibid., 21.

CHAPTER 5

TURKEY AND MISSILE TECHNOLOGY
Asymmetric Defense, Power Projection, and the Military-Industrial Complex

AARON STEIN

INTRODUCTION

Ankara has expressed a sustained interest in procuring offensive and defensive missile systems that are intended to work together to augment Turkey's capabilities to target asymmetric threats and to bolster the country's defense against a ballistic missile attack. Turkey is pursuing ballistic and cruise missiles and is eager to complement these capabilities with a robust intelligence-gathering capability that relies on space-based and unmanned systems. These systems are intended to work together to provide a better defense against regional missile proliferation.

The plans are not tied to Turkey's civilian nuclear efforts and do not appear aimed at providing Turkey with a nuclear-capable delivery system. The goals of the missile program appear to be to deepen the country's ability to target ballistic missiles before they are fired and to provide Turkish military planners with greater long-range conventional strike capabilities against a variety of targets.

Ankara is intent on producing these systems using coproduction arrangements with foreign suppliers. Thus, while the development of offensive and defensive systems is primarily aimed at addressing regional threats, there is also a substantial economic component to Turkey's military procurement and development policy. Policymakers have committed to further developing Turkey's private defense industry to increase its share of the global arms market. The strategy is tied to a larger government effort to increase research and development spending and to create high-tech Turkish products for export.

The concurrent development of cruise missiles, satellites, drones, and ballistic missile defenses is intended to provide the Turkish military with the capability to project military power beyond the country's border. This is a departure from the defensive-oriented military posture that Turkey pursued during the Cold War. These capabilities are intended to help defend against the growing threat of ballistic missiles, to solidify Ankara's position as the region's most dominant military, and to provide Turkish policymakers with a greater level of flexibility when making a decision about using military force.

As for the latter issue, Ankara has in the past had to depend on its North Atlantic Treaty Organization (NATO) allies for defense against ballistic missiles. In 1991, for example, Germany's reluctance to forward deploy aircraft as part of NATO's Allied Mobile Force–Air reinforcements raised renewed concerns about the validity of the alliance security guarantee. Moreover, the conflict exposed some of the limitations of the Turkish Armed Forces and prompted a reevaluation of doctrine. After the conflict, the armed forces opted to decrease the number of land forces in favor of a greater emphasis on modern air and missile defense systems.[1] Ankara has since had to request the forward deployment during the U.S.-led 2003 invasion of Iraq and in 2012, after Ankara grew concerned that an increasingly desperate Bashar al-Assad may launch chemical weapons at Turkish targets, or at Syrian rebel positions on the border with Turkey.[2]

THE THREAT ENVIRONMENT: THE GROWING SALIENCE OF BALLISTIC MISSILES IN THE MIDDLE EAST

In March 1985, an Iranian Scud-B short-range ballistic missile streaked into the air toward its target in Iraq. Baghdad responded with punishing air strikes and, beginning in 1988, launched some 160 extended-range Scud missiles toward Iranian cities. By that time, Iran had seized the initiative during the bloody eight-year Iran-Iraq War and was occupying Iraqi territory, including the Kurdish majority village of Halabja. After a joint Kurdish Peshmerga and Iranian Revolutionary Guards Corps offensive brought the strategically important Darbandikhan Dam into artillery range, Iraq's Ali Hassan al-Majid (Chemical Ali) used mustard gas and nerve agents to eliminate the joint Iranian-Kurdish military threat in the north. The campaign was brutal; thousands of Kurdish civilians, Peshmerga fighters, and Iranian troops were killed in the chemical attack.[3]

The growing salience of ballistic missiles, as well as the Iraqi use of chemical weapons, presented Turkish security planners with a new and vexing security challenge. Despite having remained neutral during the conflict, Turkey had tense relations with both Baghdad and Tehran and had borne the brunt of the secondary challenges posed by the conflict. For example, during the war, thousands of Kurdish refugees streamed across the Turkish border, which taxed Ankara's resources and further exacerbated Turkey's internal conflict with the insurgent Kurdistan Workers' Party (PKK).

After the collapse of the Soviet Union, Turkey emerged as the region's strongest and best-equipped military. Access to arms slowed for Soviet patrons like Iraq and Syria, while Turkey was able to benefit from the transfer of Western European and American military surplus. However, these regional imbalances prompted potential foes in the Middle East to turn to asymmetric capabilities, such as ballistic missiles and weapons of mass destruction (WMD), to compensate for their military weakness.

Thus, even though the Western bloc had become the world's most powerful military alliance, Turkey's vulnerability to asymmetric attack had grown more acute. Therefore, at the beginning of the 1990s, Turkish security planners began to pursue technologies and implement doctrines aimed at neutralizing terrorist training camps, suspected WMD sites, ballistic missiles, and other asymmetric threats.

As part of this effort, Turkey began to actively seek out ballistic missile defense, cruise missiles, remotely piloted aircraft (more commonly known as drones), and ballistic missiles. Collectively, these systems, once fully introduced, would give Turkish security planners the capability to attack missile sites both before launch and while in flight, as well as other targets such as terrorist training camps. Ankara paired these efforts with an increased emphasis on nonproliferation and global adherence to the instruments designed to prevent the spread of WMD.

MARRYING OFFENSE AND DEFENSE: TURKEY INVESTS IN MISSILES AND MISSILE DEFENSE

During the Cold War, Turkey received and was able to use American-made systems like the Honest John and the Lance missile. After the Cold War, Turkey turned its attention toward developing its own ballistic missiles using technology transfer and coproduction arrangements. In December 1995, the Turkish government announced that it had reached an agreement with Lockheed Martin for the purchase of 120 Army Tactical Missile Systems, a family of short-range, solid fuel, and satellite-guided missiles designed to attack ballistic missile launch sites, surface-to-air missiles, and command and control units.

Turkey's procurement of the system coincided with a change in Turkey's land forces doctrine, which included the movement of scores of troops away from the northern border areas to Turkey's borders with its Middle Eastern neighbors. The troop movements allowed Ankara to use the threat of military force to politically coerce neighbors and to project power into the Middle East. For example, in 1998, Turkey amassed troops on the border with Syria and demanded that Assad expel PKK leader Abdullah Öcalan from Syrian territory. As part of its military mobilization, Turkey is reported to have forward deployed its Army Tactical Missile Systems with the intention of using them to attack Syrian surface-to-air missiles and Scud missile sites, should Assad fail to comply with Turkey's demand.

Yet, absent other missile and intelligence capabilities, such a limited missile capability did not provide a robust capability to defend against ballistic missile attack. Moreover, Turkey, during its negotiations with the United States, sought to secure a favorable coproduction arrangement that would have allowed certain components of the missile system to be manufactured in Turkey. The United States refused to adhere to the coproduction terms that Turkey wanted.

Ankara then opted to partner with China Aerospace Sciences and Technology Corporation to coproduce a missile based on the B61-11 short-range ballistic missile. Turkey has since dubbed the missile the J-600 Yıldırım. After enacting a law in 1985, Turkey has strongly emphasized coproduction and offset arrangements when purchasing foreign military equipment. The law was intended to bolster Turkey's domestic military sector, which had been neglected during the Cold War.

These efforts began in the late 1970s, but the emphasis on licensing production arrangements became more prominent after the election of Turgut Özal as prime minister in 1983 and the transition from an autarkic to export-oriented economic model. Over time, Ankara has de-emphasized licensing arrangements in favor of coproduction arrangements that use a variety of Turkish-made components, known as offsets, when purchasing military technology. Turkey's insistence on such arrangements has actually led to an increase in commercial military sales abroad but has also slowed Turkey's acquisition of certain military systems and complicated negotiations with the United States.

According to F. Stephen Larrabee, "Strict U.S. restrictions on technology transfer have also caused the Turks to shun U.S. weapon systems and turn to non-U.S. manufacturers, such as Israel and South Korea."[4] Moreover, in the U.S. Congress, numerous concerns have been raised about Turkey's human rights record, which has led to hesitancy about approving arms exports. These restrictions and concerns have prompted Ankara to look to other suppliers for military equipment.

In the late 1990s, for example, Ankara and Israel cooperated quite closely on weapons issues and entered into offset agreements for the sale of drones and cruise missiles. Turkey's turn to Israel for advanced weapons did not come about because of any deep-seated political affinity for the Jewish state but was driven by Israeli willingness to overlook Turkey's human rights issues and the opportunity to import high-tech weapon systems under favorable arrangements for Turkish defense industries.

In 1997, for example, Israel and Turkey engaged in a series of discussions for the sale of Israel's Arrow II missile interceptor and accompanying

Green Pine radar system.[5] Turkey had coveted the Arrow II system because it is specifically built to engage Scud-type threats. The system is designed to track and destroy "missiles with ranges of up to 1,500 km and to intercept incoming missiles at a much higher altitude—50 km—than the American Patriot system."[6] Yet, in order for Israel to export the system, it must first receive permission from the United States. And despite close Turkish-American ties, the United States government objected[7] to the transfer of the system, over concerns that the Arrow II is classified as a category 1 system under the Missile Technology Control Regime.[8]

The regime is a voluntary arrangement of 33 countries, including the United States and Turkey (Israel is not a member, but claims to follow its export control provisions), aimed at preventing the spread of missiles and related technologies capable of carrying a 500 kilogram payload at least 300 kilometers. Participating states are expected to maintain a strong presumption of denial on exports of category 1 items, which includes the transfer of complete missile systems, rocket engines, and other critical subsystems.[9]

In spite of U.S. objections to the Arrow's export, Turkish security officials continued to pursue missile defense as a top priority.[10] In 2000, for example, the Turkish Armed Forces announced their intention to procure a two-tiered missile defense system that would use the U.S. THAAD system (terminal high altitude area defense) and/or the Arrow and Patriot Pac-3 systems. Ankara was interested in cooperating on the Arrow project, so as to benefit from technology transfer.[11] Ankara was forced to cancel the planned missile defense deal in 2001, however, after a severe financial crisis caused the armed forces to scale back their missile defense ambitions.

Turkish-Israeli cooperation in cruise missile technology began in 1996, shortly after the signing of an agreement on "Military Training and Cooperation."[12] In 1997, Turkish and Israeli officials agreed to the sale of 46 air-launched Popeye-1 land attack cruise missiles. The missile has a range of 80 kilometers and is designed for standoff strikes. In addition, the two sides agreed to form a joint venture between Israel's Rafael and Turkey's Roketsan for coproduction of the Popeye-2.[13] The coproduction deal, however, never came to fruition. Until that point, Turkey had imported only the American anti-ship Harpoon cruise missile.

Turkey's indigenous effort to develop a cruise missile began in 2006. The first prototype, known as the standoff munition, or SOM, began wind tunnel testing in 2008, and the first live fire test took place in 2011.[14] The missile is reported to have flown 185 kilometers and struck its pre-programmed target. The SOM is designed to be carried by the F-4 and the F-16D. Turkey's Roketsan has also signed an agreement with Lockheed Martin to collaborate on the development of the SOM-J, a version with folding tail fins, so that it can be carried internally by the F-35.[15] Ultimately, Turkey intends to deploy three SOM variants. The SOM A will have a high-explosive unitary warhead, the SOM B will carry a similar warhead, and the SOM B-2 will carry a penetrating warhead. Thus far, only the SOM A and SOM B have undergone testing.

The SOM is powered by the French-built Microturbo TR-40 turbojet engine. In 2011, Turkey awarded a contract to Kale Havacılık for the local production of a replacement turbojet engine. The engine is expected to combine the best attributes of the TR-40 with systems thought to be similar to the Teledyne CAE J402 series used to power the U.S. Harpoon.[16] According to Murad Bayar, the former head of Turkey's defense procure-ment agency, the SOM engine project is expected to be completed in late 2015.[17]

Moreover, in 2012, the TÜBİTAK SAGE director, Yücel Altınbaşak, indicated that the range could be extended to 500 kilometers if the mis-sile's payload weight were decreased.[18] All three variants are designed to strike such targets as surface-to-air missile sites, exposed aircraft, command and control facilities, and aircraft hangars from standoff range with accu-racy and precision.[19] Once completed, the missile will provide the Turkish Armed Forces with the capability to strike targets with accuracy from long distances. The missile could theoretically be used in a conflict on the island of Cyprus, in an attack on command-and-control facilities in neighboring countries Iran and Syria, or in a strike on mobile missile launchers.

Ankara intends to pair these offensive capabilities with a locally pro-duced missile defense system. Turkey's Aselsan and Roketsan have been awarded a $1 billion contract to develop low- and medium-range missile defense systems.[20] According to Burak Bekdil, Aselsan is reported to be developing "all radar, fire control, command-and-control and com-munication systems for both the low-level and medium-altitude compo-nents of the program," including the infrared sensors and data links for the missile's warhead.[21] The program is ambitious, and it is unlikely that it will be deployed quickly. Nevertheless, it indicates the type of technologies that Ankara hopes to produce in the coming decades.

Turkey's indigenous missile defense program has moved in parallel to its interest in procuring a long-range missile defense system from a foreign supplier. In 2008, Turkey first put forward a tender for long-range air and missile defense systems. In 2009, Ankara was reported to be considering bids from Raytheon and Lockheed Martin in the United States for the Patriot air and missile defense system; Russia's Rosoboronexport S-300, with the possibility of negotiation for the S-400; China's Precision Machinery Import Export Corporation (CPMIEC) HQ9 (the export version is called the FD-2000); and the Italian-French Eurosam, SAMP/T Aster 30. Turkey had initially sought to purchase an off-the-shelf system but changed the terms of the tender in January 2013 to a coproduction arrangement.

Ankara eventually opted for China's HQ-9, even though the system is not interoperable with the sensors—including the X-band radar deployed in southern Turkey—that the United States and other European states are deploying as part of the NATO missile defense system. Turkish officials maintain that they chose the Chinese system because the HQ-9 was the least expensive system, CPMIEC offered the most favorable technology transfer terms, and the Chinese firm was willing to coproduce the entire system with Turkey. In addition, Ankara is likely to receive an indirect offset in the form of a technology park built on the outskirts of Istanbul.

Turkey likely intends for the HQ-9 to be a stopgap system to be used until its own missile defense system is fully developed. Ankara is developing all of the ancillary systems a state would need to support a national missile defense system—including components for a sea-based missile defense system—though all of the technologies in development are in their infancy and are unlikely to provide a capable defense against missile attack in the medium to long term. Ankara may seek to replicate some of the systems used in the HQ-9 for its own missile defense needs. And, once fully developed, Turkey is likely to seek to export the system. However, given the immense difficulties in developing missile defense systems, it is unclear if Turkey will ever be able to field a capable system that it can also export.

Ankara has yet to finalize the purchase agreement with China, and reports suggest that a combination of diplomatic pressure from Turkey's Western allies, as well as concerns about interoperability, may result in Ankara reissuing the T-Loramids tender.[22] Turkey has indicated that it is prepared to reevaluate its initial decision and, to this end, has begun parallel talks with MBDA about the SAMP/T system. Ahmet Davutoğlu has previously said that the initial decision was made on three criteria: "joint production, the time of delivery and price."[23] Davutoğlu, who was foreign minister when he said that and is now the prime minister, has made clear that both MBDA and Raytheon/Lockheed must improve in all three areas to win the tender.

The United States, however, maintains strict restrictions on the transfer of military technology. In the past, U.S. export policies have prompted Turkey to turn to Israeli and South Korean firms for other technology such as tanks, drones, and missiles.[24] Moving forward, it is unlikely that Raytheon will ever be able to meet Turkey's transfer demands.

Turkey has had more success with European suppliers. Most notably, Ankara spurned American helicopter suppliers in favor of a European supplier after a tumultuous tender process finally resulted in AgustaWestland agreeing to coproduce an attack helicopter in Turkey. Similarly, MBDA signed an agreement with Turkish subcontractors in November 2010 that would have allowed the transfer of certain design information for missile defense components to Turkish firms, should Ankara have chosen the Samp/T.[25] The agreements, according to Turkey's Anadolu Agency, would have "offered Turkey complete freedom of technology transfer in case of cooperation."[26]

If Ankara does opt to choose a different supplier, it would appear that MBDA has a greater chance of winning the tender than the Raytheon/Lockheed consortium. Moreover, it suggests that Ankara elevated price over interoperability when it first chose CPMIEC's HQ-9. Yet, in doing so, Ankara was ill-prepared for the criticism from its NATO allies. In addition, Turkish policymakers underestimated defense-related difficulties stemming from the decision to select a supplier on which the United States Treasury has imposed sanctions. U.S. financial firms, for example, have since refused to assist Aselsan with its initial public offering, over concerns that doing so could violate U.S. law.[27]

The cost associated with the procurement of the system is likely to continue to play a large role in the final decision. Ankara has a relatively small defense acquisition budget, which underscores the importance of price in its decisionmaking. In total, Turkey's sustained interest in developing ballistic and cruise missiles suggests that Turkish security planners are intent on having the capability to strike targets in the region from standoff ranges. Turkey intends to pair its offensive capabilities with its burgeoning defensive capabilities to put in place a robust defense against ballistic missiles. However, it is unwilling to seriously compromise on its coproduction demands, which will certainly delay the procurement of a credible system in the near future. Nevertheless, this represents a shift in Turkish military doctrine away from a defensive-oriented posture, to a doctrine that includes scenarios in which Turkey could use precision strikes in neighboring countries.

INTELLIGENCE DEMANDS: TURKEY SEEKS GREATER INDEPENDENCE

As part of this shift, Turkey is developing greater intelligence, satellite, and reconnaissance capabilities. These systems are intended to supply targeting information for Turkey's planned missile systems and, critically, provide Ankara with more independence from U.S.-supplied intelligence. Turkey intends to coproduce and launch up to 16 satellites by 2023—the centennial of the founding of the Turkish Republic. The military satellites will be outfitted with systems for reconnaissance and communications. Moreover, one is expected to include infrared sensors for ballistic missile launch detection.[28]

To support these plans, Ankara announced plans to develop an independent satellite launch capability.[29] In February 2012, the head of TÜBİTAK SAGE announced that Turkey had an ambitious plan to build a ballistic missile with a 2,500 kilometer range within two years.[30] Shortly after the missile announcement, the defense procurement agency also announced that it would begin work on a satellite launch vehicle[31] and construct a satellite launch pad to support Turkey's ambitious plans for space.[32] The plan for both systems remains opaque, and little information has been released about the status of either program. As a result, it is unclear if the satellite launch vehicle is intended to serve as the design platform for the ballistic missile, or vice versa.

Ankara has also indicated that it intends to use its satellite launch vehicle and its future satellite industry to help establish Turkey as a major exporter of space-related technology. Thus, as with other military programs, the government is eager to use the technology to enhance the Turkish economy. Yet, as with the missile defense plan, Ankara is unlikely to be able to realize its very ambitious goals in the time frame allotted. While Turkey is working with European and Japanese suppliers for the development of satellites, it is unlikely that Ankara will be able to successfully develop and test a rocket capable of placing satellites in orbit within two years.

As for the military use of ballistic missiles, the armed forces and civilian leadership have not indicated why they would need a ballistic missile with a range of 2,500 kilometers or what role the missile would play in Turkey's defense policy. However, other states that employ conventional ballistic missiles are known to use them to target command-and-control sites, large military bases, and ships at sea.

Ankara intends to network its space-based systems with its recently procured airborne warning and control system surveillance planes and the Anka drone using a locally developed software system dubbed Herikks, or Skywatcher. The systems are designed to work in concert to increase Turkish intelligence capabilities. In the past, Ankara has used its current fleet of drones to provide targeting information for field artillery units and to direct F-16 airstrikes. Thus, it is likely that once these systems come online, Turkey will use drone intelligence for similar operations, as well as for its cruise missiles.

The Anka is part of a larger effort to acquire greater precision strike capabilities. Ankara intends to pair the Anka with Roketsan's 8 kilometer range Mızrak-U anti-tank missile. Ankara's precision-strike investments necessitate greater intelligence, satellite, and reconnaissance capabilities. In turn, the Anka is expected to improve Turkey's ability to identify targets, as well as carry out hunter-killer missions to shorten the kill-chain.[33] Turkey began work on the Anka's airframe design, software, and communication sub-systems in the early 1990s, after gaining experience in drone operation from the import of the American-made Gnat. The initial design work culminated in a 2004 government tender to design and develop a medium-altitude, long-endurance drone.[34] The contract for the development and production of the drone was awarded to TAI. The drone has since been named the Anka and is reported to be able to fly at an altitude of up to 30,000 feet for up to twenty-four hours.[35]

The Anka uses a VHF/UHF link. This means that Turkey's data-transfer burden is less demanding than the American Predator, but it also limits the aircraft's range to line of sight.[36] Turkey intends to outfit the Anka with satellite communications and, perhaps after the launch of its own communications satellites, may seek to use the technology to pilot the aircraft and to transmit data collected. Yet, like other emerging drone nations, Ankara is likely to face a number of challenges ranging from the cost of drone development to the strains the extensive use of drones will place on intelligence services as it seeks to increase its use of unmanned systems. Turkey claims that the aircraft's testing ended in January 2013, but production has lagged.[37]

During testing, the drone repeatedly crashed, raising numerous questions about its reliability, as well as the potential procurement costs, should the drone continue to crash once it begins service with the Turkish Armed Forces.[38] Moreover, Turkey has faced some problems procuring engines for the Anka. Initially, Turkey intended to purchase ten diesel engines from Germany's Thielert. However, after China's AVIC International purchased the struggling European firm, Turkey was forced to look elsewhere.[39] Ankara eventually opted to produce its own engines. According to Bayar, Turkey's procurement agency signed an agreement with state-owned Tusas Motor Sanayi to produce five prototype engines for the Anka. The design, production, and certification tests are expected to be completed by 2017.[40]

In parallel to its indigenous efforts, Turkey has imported unarmed surveillance drones from the United States and Israel. The Turkish military was the first foreign purchaser of the Gnat, the platform that spawned today's Predator. Turkey imported 24 Gnats, including two control stations, between 1994 and 1996.[41] Later, in 2005, Turkey purchased ten Heron medium-altitude, long-endurance drones from Israel Aerospace Industries. Yet, in this case, Turkey's offset demands delayed the introduction of the Israeli-made drone.

Ankara had conditioned the deal on the Heron's use of Aselsan's Aselfir 300T turret, which reportedly weighs 50 pounds more than the Heron's normal MOSP turret, or the similar 155-pound AAS-52 system carried by the Predator.[42] The extra weight reportedly limited the Heron's altitude and endurance, which then brought it below the original tender's altitude and endurance thresholds. Thus, in order to meet Ankara's immediate need for drones, Turkey agreed in 2007 to temporarily lease three Aerostar drones and one stock Heron drone. TAI eventually delivered the ten Herons in 2010, and the drones are used extensively in the Kurdish-majority southeast.

MISSILE DEFENSE AND DETERRENCE: TURKEY DE-EMPHASIZES NUCLEAR WEAPONS

Ankara's embrace of missile defense and precision strike has moved in parallel to its de-emphasis of nuclear weapons for its security. Turkey has hosted U.S. tactical nuclear weapons since 1959. During the Cold War, Ankara embraced the use of nuclear weapons, and even opted to delay its signing of the Treaty on the Non-Proliferation of Nuclear Weapons over concerns that doing so could prevent the military from being authorized to detonate the weapons on Turkish territory.[43] Yet, as of 2005, Turkey turned down an opportunity to station a permanent U.S. dual-capable fighter wing at İncirlik Air Base—the only base in Turkey that continues to host tactical nuclear weapons.

As of 2010, Turkey is estimated to host 60–70 B61 gravity bombs. Fifty of these bombs are still slated for delivery via U.S. aircraft. The other ten to twenty are reserved for delivery by Turkish F-16s. However, General Ergin Celasin, commander of the Turkish Air Force until 2001, is on record saying that "Turkey's role in NATO's nuclear contingency plans came to an end with the withdrawal of nuclear weapons in the 1990s from the Air Force units that were deployed in several air bases in Turkey."[44]

Yet, according to Hans Kristensen, until Turkey receives the F-35, "approximately 30 F-16C/D Block 50s are scheduled to receive a 'stopgap' upgrade to make them capable of carrying the new B61-12 bomb that will replace the B61-3/4 beginning in 2019."[45] Thus, it is widely assumed that Turkish aircraft have the capability to deliver nuclear weapons, but its pilots are no longer certified to do so. In any case, Turkey has adopted a unique nuclear posture, whereby both the 52nd U.S. fighter wing and either the 4th fighter wing stationed at Akıncı Air Base or the 9th Turkish fighter wing stationed at Balıkesir Air Base would have to fly to İncirlik and be loaded with nuclear weapons before flying on to their targets.

More broadly, Ankara's nuclear posture reflects the 2010 U.S. Nuclear Posture Review. Like the United States, Ankara has embraced the diminished role of nuclear weapons but has taken numerous steps to ensure that it develops a capable missile defense system to defend against WMD threats. Nevertheless, Turkish officials continue to espouse a worldview that embraces disarmament, while also recognizing the perceived necessity of maintaining a credible NATO deterrent, so long as nuclear weapons exist.[46] Ankara's long-term policy, therefore, is to blend its conventional capabilities with forward-deployed NATO assets to deter potential WMD threats. Ankara's conventional weapons are designed to work in tandem with NATO doctrine, which ultimately rests on the threat of nuclear weapon use to deter WMD armed states. Turkey, therefore, has not shown any willingness to support the withdrawal of U.S. nuclear weapons from European air bases, even though it now views the B61 as a "political weapon" intended to promote alliance solidarity.

CONCLUSION: AMBITIOUS PLANS AND A SUSTAINED COMMITMENT

Turkey's missile programs are largely aimed at increasing the military's capabilities to target asymmetric threats. The original impetus for the program came about after the Cold War and coincided with a change in Turkey's military doctrine. The new doctrine calls for the development of a defense against a range of threats, including the proliferation of ballistic missiles and WMD in the Middle East.

These efforts have been slowed by demanding coproduction and offset arrangements with potential foreign suppliers. Nevertheless, Turkish officials remain committed to pursuing the arms procurement policy that was first codified in 1985. The programs, therefore, are also tied to the overall effort to develop a private defense industry and to eventually turn Turkey into a net military exporter. The programs are ambitious and unlikely to be completed on time. Nevertheless, they represent the ambition of policymakers and signal a sustained commitment to deepening Turkey's technological base. Moving forward, Ankara is certainly going to pursue a similar strategy as it seeks to realize its goals.

NOTES

1 Ian O. Lesser, *Bridge or Barrier: Turkey and the West After the Cold War* (Santa Monica, Calif.: RAND Corporation, 1992), 30–32.

2 Shashank Joshi and Aaron Stein, "Turkey's Troubles in Syria," *RUSI Journal* 158, no. 1 (Spring 2013): 27–39.

3 David Segal, "The Iran-Iraq War: A Military Analysis," *Foreign Affairs* 66, no. 5 (Summer, 1988): 946–63.

4 F. Stephen Larrabee, *Troubled Partnership: U.S.-Turkey Relations in an Era of Global Geopolitical Change* (Santa Monica, Calif.: RAND Corporation, 2010), 78.

5 Arieh O'Sullivan, "Defense Ties With Turkey Bolstered," *Jerusalem Post*, December 9, 1997.

6 Dennis Gormley, *Missile Contagion: Cruise Missile Proliferation and the Threat to International Security* (Annapolis, Md.: Naval Institute Press, 2008), 29.

7 Şebnem Udum, "Missile Proliferation in the Middle East: Turkey and Missile Defense," *Turkish Studies* 4, no. 3 (Autumn 2003): 86–87.

8 Wade Boese, "U.S.-Israeli Policy for Exporting Arrow Missile Undecided," Arms Control Association, September 2002, www.armscontrol.org/print/1097.

9 Missile Technology Control Regime, "Guidelines for Sensitive Missile-Relevant Transfers," www.mtcr.info/english/guidelines.html.

10 First Lieutenant Guray Al, "Turkey's Response to Threats of Weapons of Mass Destruction," MA thesis, Naval Postgraduate School, 2001, 117–18.

11 Lale Sariibrahimoğlu, "Army Defines New Missile Strategy," *Hürriyet Daily News*, February 9, 2000, http://web.hurriyetdailynews.com/army-defines-new-missile-strategy.aspx?pageID=438&n=army-defines-new-missile-strategy-2000-02-09.

12 William Hale, *Turkish Foreign Policy Since 1774*, 3rd ed. (London: Routledge, 2013), 227.

13 Philip Robins, *Suits and Uniforms: Turkish Foreign Policy Since the Cold War* (Seattle: University of Washington Press, 2003), 201.

14 "Turkish Cruise Missile Design Breaks Cover," *Flight Global*, September 14, 2011, www.flightglobal.com/news/articles/dsei-turkish-cruise-missile-design-breaks-cover-362026.

15 Burak Bekdil, "Lockheed Martin Signs Deal With Turkish Missile Maker," *Defense News*, October 27, 2014, www.defensenews.com/apps/pbcs.dll/article?AID=2014310270014.

16 "KALE Aero Will Develop Indigenous Turbojet Engine for SOM ALCM," *Monch Turkiye*, www.monch.com.tr/index.php?option=com_content&task=view&id=158.

17 Burak Ege Bekdil, "Interview With Murad Bayar: Head of SSM, Turkey's Defense Procurement Agency," *Defense News*, November 25, 2013, www.defensenews.com/article/20131125/DEFREG01/311250016?utm_source=twitterfeed&utm_medium=twitter.

18 Umit Enginsoy, "Turkey Aims to Increase Ballistic Missile Ranges," *Hürriyet Daily News*, February 1, 2012, www.hurriyetdailynews.com/turkey-aims-to-increase-ballistic-missile-ranges.aspx?pageID=238&nID=12731&NewsCatID=345.

19 TUBITAK SAGE, "SOM, Modular Stand-Off Missile," www.sage.tubitak.gov.tr/en/urunler/stand-missile-som.

20 Burak Ege Bekdil, "Aselsan Wins $1B Turkish Air Defense Contract," *Defense News*, June 23, 2011, www.defensenews.com/article/20110623/DEFSECT01/106230308/ Aselsan-%20Wins-1B-Turkish-Air-Defense-Contract.

21 Ibid.

22 Burak Ege Bekdil, "Turkey Seeks New Extension in Air Defense Contract," *Defense News*, December 22, 2014, www.defensenews.com/story/defense/air-space/2014/12/22/turkey-seeks-new-extension-in-air-defense-contract/20757225.

23 "Turkey Says Open to Alternatives to Chinese Missile Defense System," Reuters, February 2, 2014, www.reuters.com/article/2014/02/02/us-turkey-china-defence-idUSBREA1108L20140202.

24 Larrabee, *Troubled Partnership: U.S.-Turkey Relations in an Era of Global Geopolitical Change*, 77–79.

25 "Italy-Turkey: MBDA Signs Framework Deal With Turkish Firms," ANSA English Corporate Service, November 11, 2010.

26 "Arms Firm MBDA Hopeful to Build Turkey's Air Defense," Anadolu Agency, December 6, 2010.

27 Cengizhan Çatal, "Aselsan Eyes Goldman Sachs After Chinese Missile Row," *Hürriyet Daily News*, December 13, 2013, www.hurriyetdailynews.com/aselsan-eyes-goldman-sachs-after-chinese-missile-row.aspx?PageID=238&NID=59484&NewsCatID=345.

28 Burak Ege Bekdil, "Turkey Plots Path Towards Space Command," *Defense News*, April 9, 2013, www.defensenews.com/article/20130409/DEFREG01/304090011/ Turkey-Plots-Path-Toward-Space-Command.

29 Ibid.

30 Umit Enginsoy, "Turkey Aims to Increase Ballistic Missile Ranges," *Hürriyet Daily News*, February 1, 2012, www.hurriyetdailynews.com/turkey-aims-to-increase-ballistic-missile-ranges.aspx?pageID=238&nID=12731&NewsCatID=345.

31 "Turkey Going Ballistic, Eyes Space-Launched Vehicle," *World Tribune*, August 12, 2012, www.worldtribune.com/2012/08/12/turkey-going-ballistic-eyes-space-launched-vehicle.

32 Emre Soncan, "Turkey Plans to Build Own Satellite Launch Base for Peaceful Purposes," *Today's Zaman*, September 6, 2012, www.todayszaman.com/newsDetail_getNewsById.action?newsId=291583.

33 Burak Ege Bekdil, "For Turkey, Precision Is Maximum Lethality, Minimum Cost," *Defense News*, January 21 2014, www.defensenews.com/article/20140121/ DEFREG01/301200037/For-Turkey-Precision-Maximum-Lethality-Minimum-Cost.

34 Aaron Stein, "Turkey's Missile Programs: A Work in Progress," EDAM Policy Brief, Center for Economics and Foreign Policy Studies, January 2013, http://edam.org.tr/ disarmament/EN/documents/Turkey%20Missile%20Programs.pdf.

35 Turkish Aerospace Systems, ANKA Medium Altitude Long Endurance UAV System, www.tai.com.tr/en/project/anka-medium-altitude-long-endurance-uav-system.

36 Turkish Aerospace Industries, "ANKA Successfully Completes Acceptance Tests," January 25, 2013, www.tai.com.tr/en/basin-bultenleri/anka-successfully-completes-acceptance-tests.

37 Emre Soncan, "Mass Production of Turkey's First National UAV Imminent," *Today's Zaman*, January 24, 2013, www.todayszaman.com/news-305039-mass-production-of-turkeys-first-national-uav-imminent.html.

38 "Unmanned Air Vehicle ANKA Fails Final Test," *Hürriyet Daily News*, September 29, 2012, www.hurriyetdailynews.com/unmanned-air-vehicle-anka-fails-final-test.aspx?pageID=238&nid=31278.

39 Burak Bekdil, "China's Buying of German Firm Risk for Turkish UAV," *Hürriyet Daily News*, September 17, 2013, www.hurriyetdailynews.com/chinas-buying-of-german-firm-risk-for-turkish-uav.aspx?pageID=238&nID=54563&NewsCatID=483.

40 Bekdil, "Interview With Murad Bayar."

41 See SIPRI Arms Transfers Database, "Global Transfers of Major Conventional Weapons: Sorted by Supplier. Deals With Deliveries or Orders Made for Year Range 1990 to 2011."

42 "Israeli Manufacturers' Turkish UAV Contract," *Defence Industry Daily*, December 22, 2011, www.defenseindustrydaily.com/israeli-manufacturers-win-150m-turkish-uav-contract-updated-0389.

43 Aaron Stein, "Turkey and the Backpack Bombs," WMD Junction, Center for Nonproliferation Studies, February 6, 2014, http://wmdjunction.com/140206_turkey_and_adms.htm.

44 See telephone interview with General Ergin Celasin (ret.), former commander of the Turkish Air Force, February 15, 2010, Ankara, cited in Mustafa Kibaroğlu, "Turkey and Shared Responsibilities," in *Shared Responsibilities for Nuclear Disarmament: A Global Debate* (Cambridge, Mass.: American Academy of Arts and Sciences, 2010), 27.

45 Hans Kristensen, *Non-Strategic Nuclear Weapons* (Washington, D.C.: Federation of American Scientists, May 2012), www.fas.org/_docs/Non_Strategic_Nuclear_Weapons-lr.pdf.

46 Mustafa Kibaroğlu, "A Warm Reception: Turkey (Mostly) Embraces Obama's Nuclear Posture," *Nonproliferation Review* 18, no. 1 (March 2011).

CHAPTER 6

TURKEY, THE NONPROLIFERATION TREATY, AND THE NUCLEAR SUPPLIERS GROUP

MARK HIBBS

The Turkish Republic is a member of two arrangements that largely define its foreign policies in the nuclear energy sphere: the Treaty on the Non-Proliferation of Nuclear Weapons (NPT) and the Nuclear Suppliers Group (NSG). When Turkey joined both the NPT and the NSG it did so because it concluded that membership was consistent with the country's values, aspirations, and international commitments. At the time they were made, these decisions did not excite public debate or controversy, and Turkish leaders had little to say to explain their actions.

Over time, Turkey's views of both the NPT and the NSG have evolved and the country has developed narratives to articulate Turkey's national interest concerning both nonproliferation and nuclear trade. As was the case when Turkey joined the NPT more than three decades ago, Turkey's policies in these areas today express support for nuclear disarmament, nonproliferation, and the deployment of nuclear technology for peaceful purposes.

TURKEY AND THE NPT

Turkey signed the NPT on January 28, 1969. It could have been anticipated then that Ankara would follow by quickly ratifying the NPT to bring it into force in Turkey for three basic reasons: the role of the North Atlantic Treaty Organization (NATO) in Turkey's national security; Ankara's relationships with the United States and the European Economic Community (EEC); and Turkey's desire for peaceful nuclear cooperation.

In 1969, Turkey had been a member of NATO for seventeen years, and it was formally protected by the alliance's security guarantees including Article 5 of the North Atlantic Treaty committing each member state to consider an attack against one state to be an attack against all NATO members. Under these circumstances Turkey's leaders took for granted that Turkey shouldn't need nuclear weapons.

Since 1964, Turkey had also been an associate member of the EEC (a forerunner of the European Union), and its agreement with Brussels included a road map that would permit Ankara twenty-three years thereafter to apply for full membership. In the meantime, the EEC had replaced the United States as Turkey's most important trading partner, and there were grounds for optimism that Turkish-EEC relations would continue to intensify.[1] All five EEC states without nuclear weapons in 1969 signed the NPT and then ratified the treaty simultaneously, underscoring their common interest in taking this action.

Twelve years before it signed the NPT, Turkey had joined the International Atomic Energy Agency (IAEA) to support Ankara's ambition to develop the peaceful atom with assistance from foreign partners. Two years later, in 1959, under Washington's "Atoms for Peace" initiative, and in the same year that NATO deployed U.S. nuclear weapons on

Turkish territory, the United States agreed to supply Turkey with its first nuclear reactor. This unit, TR-1, was promptly set up in Istanbul by the American Machine Foundry Company, and the reactor began operating in 1962.[2] Beginning five years later, the United States urged Turkey to join what would become the NPT.

But after Turkey signed the NPT in 1969, it did not ratify the treaty for eleven years. All but two of NATO's other thirteen members in 1969—France and Portugal—had ratified the NPT by 1975. The delay in Turkey's ratification until 1980 has prompted speculation that in 1969 Turkey was not prepared to forswear the option to obtain nuclear weapons in the future.[3]

Senior Turkish officials interviewed in 2012 and 2013 denied that Turkey had ever considered developing and deploying nuclear arms. The eleven-year NPT ratification delay, one said, was mainly caused by an extended domestic political crisis unleashed during the late 1960s. For months prior to and after Turkey's NPT signature, opposition to the country's civilian government escalated into severe violence and seething anti-American demonstrations. In response to the civil disorder, military leaders hatched plots that ultimately culminated in a military ultimatum to the government in 1971.[4] The political crisis continued for a decade. "[T]hirteen weak coalition governments [that] had swapped power amid growing political instability led inexorably to the more sweeping military coup"[5] of September 1980. Four months before that coup, the last civilian coalition government of Turkey pushed NPT ratification through parliament.

According to the above-cited Turkish official, after Turkey's NPT signature neither civilian nor military leaders were inclined to expedite ratification since during the 1960s and early 1970s the United States had angered Turkey by agreeing secretly with the Soviet Union during the 1962 Cuban Missile Crisis to withdraw nuclear weapons from Turkey, by pressuring Ankara in 1963 not to intervene militarily on Cyprus, and by imposing an arms embargo in 1974 after Turkey invaded Cyprus. The political risk that anti-U.S. sentiment in Turkey represented throughout the 1970s, according to a former U.S. diplomat, discouraged Turkish leaders from making decisions that could be framed by opponents as capitulating to Washington. That was especially the case for Turkey's NPT membership, he said, because Turkish leaders did not assign it a high priority.

When Turkey finally ratified the NPT, the government's rationale was expressed in a statement that Ankara filed together with its ratification instruments to depositary governments in April 1980. The statement expresses views on three key issue areas: nuclear disarmament, nonproliferation, and peaceful use of nuclear energy. These views have prevailed in Turkish politics, and they are broadly consistent with current Turkish positions. A senior Turkish official said in an interview that Turkey—then as now—holds the view that nonproliferation, disarmament, and peaceful use of nuclear energy constitute the NPT's "three pillars," an idea that is axiomatic for many NPT parties.

If Turkey initially did not consider the NPT very important to Turkey's security, by 1980 that situation may have been changing. Turkey's ratification statement called the NPT "the most important multilateral arms control agreement yet concluded"—an expression of confidence that in hindsight seems hyperbolic but at the time reflected increasing Turkish concern about the need for nuclear-weapons reductions. "By reducing the danger of a nuclear war," Turkey said, the NPT "greatly contributes to the process of détente, international security and disarmament."

Today, many states voice aspirations consistent with Turkey's 1980 statement, including its assertion that "the cessation of the arms race ... can only be realized through the conclusion of a treaty on general and complete disarmament under strict and effective international control," and the reminder that the nuclear-weapon states in the treaty had incurred "non-proliferation obligations ... under relevant paragraphs of the preamble and Article VI of the treaty." Turkey's references in the document to the threat of proliferation and the importance of security assurances also proved to be long-standing. Turkey mentioned that "proliferation of all kinds must be halted and measures must be taken to meet adequately the security requirements of non-nuclear weapons states. Continuing absence of assurances might have consequences that could undermine the objectives and provisions of the treaty." Finally, the statement noted that because Turkey aspired to develop nuclear power, it sought to "cooperate with technologically advanced states on a non-discriminatory basis in the field of nuclear research and development as well as in nuclear energy production," and in the same breath Ankara asserted that "measures developed or to be developed at the national and international levels to ensure ... non-proliferation ... should in no case restrict the non-nuclear weapons states in their options for the application of nuclear energy for peaceful purposes."[6] These 1980 Turkish assertions are consistent with the mainstream NPT rhetoric of many states today.

TURKEY'S NPT EVOLUTION

All of these positions expressed in the 1980 NPT ratification statement were frequently reiterated by Turkish diplomats during the decades that followed. Over time, to the extent that nonproliferation and nuclear energy have become more important factors in Turkish foreign policy considerations, Turkey's positions have evolved and have become in some areas more clearly formulated. In some cases, contradictions arose between Turkey's NPT-based aspirations and other policy goals, and gaps were exposed between Turkey's aspirations and the political will of the central government to realize and enforce them.

Nuclear Weapons and Nuclear Disarmament

Turkish policymaking and diplomacy related to the NPT have become increasingly bedeviled by tensions between Turkey's support for nuclear disarmament and its commitments to the NATO alliance. This problem appeared on the horizon even before Turkey signed the NPT. A U.S. State Department document from 1967—eight years after U.S. nuclear weapons were first deployed on Turkish territory—spelled out that prior to signing the NPT, Turkey wanted to know whether "predelegation of the authority to fire nuclear weapons would be permissible under the treaty."[7]

Ever since Turkey ratified the treaty, Ankara's NPT diplomacy has expressed rhetorical support for nuclear disarmament while, outside NPT venues, Turkey has embraced NATO's nuclear-armed guarantees and deterrence posture as the cornerstone of its security policy. It may be true that Turkey's support for disarmament is logical in the NATO context because Turks are weary of being hostage to Russia's Cold War–era nuclear weapons.[8] But Ankara's concerns about the perceived nuclear threats it faces, including potential threats, have long overshadowed Turkey's interest in ridding the world of nuclear weapons.

Turkey remains one of a handful of NATO countries that deploy forward-based nonstrategic nuclear weapons on their territories. It is not obvious that Ankara anytime soon will on its own undertake any initiative to remove these weapons from Turkey. Unless it does withdraw them—perhaps in consort with the removal of such systems in other NATO member states—it will be difficult for Turkey to assume a more credible and prominent nuclear disarmament profile in its NPT diplomacy.

Turkey and its NATO allies joined the NPT at a time when the Cold War was intensifying and the shadow of mutual assured destruction was growing on the Soviet periphery. So long as Turkish leaders perceived the Soviet Union to be Turkey's greatest threat, Ankara's commitment to disarmament would remain underdeveloped and marginal. This was especially the case because Russia is a near neighbor and has been an enemy for much of Turkey's modern history. The two countries have gone to war numerous times since the eighteenth century, and Turkey has experienced Russian incursions in the Black Sea, the Balkans, and the Caucasus.[9]

The end of the Cold War deflated Turkey's concern about meeting Soviet nuclear-weapon threats. Détente, followed by the collapse of the Soviet Union, also permitted Turkey more freedom to explore new diplomatic opportunities.

But while Turkey joined most other states in supporting U.S.-Soviet nuclear arms reductions, Turkish leaders during the 1990s expressed concern that the end of bipolarity would destabilize Turkey's neighborhood including the Middle East. At the same time, the end of the Cold War raised questions about Turkey's future relations with the United States and NATO. The U.S.-led invasions of Iraq in 1991 and 2003 caused serious diplomatic problems between Ankara and Washington. During the 2003 Iraq War, "elite opinion began to follow the general public with rising nationalist sentiment, growing skepticism over US motives and memories of the cost incurred during the Gulf War in 1991."[10]

Indeed, Turkey over the years accumulated a series of grievances that, in its view, cast reasonable doubt about whether NATO would meet its collective defense commitments to Turkey in a crisis, especially one involving a nuclear-armed adversary.[11] In both Iraq wars some NATO allies would not deploy early-warning radar and missile defense systems in Turkey and would not agree to consultations that Turkey had sought under the North Atlantic Treaty.

As Turkish relations with the United States seemed to cool in the post–Cold War era, Ankara responded by looking for diplomatic opportunities in its region, including with Russia, and by generally moving beyond viewing its options through the prism of its relations with Western countries. This reevaluation was abetted by opposition in the European Union (EU) since the late 1990s to efforts by Turkey to join the EU.

Turkey's rebalancing was considerably encouraged by the rise of the Justice and Development Party (AKP), which since 2002 has governed Turkey with comfortable parliamentary majorities. The foreign policy of this Islamist party under Recep Tayyip Erdoğan, the prime minister from 2003 to 2014 and now the president, has been less informed by U.S. and Western views, continuing developments in Turkish diplomacy that got under way beginning in the 1980s.[12]

The AKP has been committed to a substantial reworking of Turkey's policies toward states in the Middle East, prompting efforts by Turkey to intervene to try to negotiate a partial solution to the Iran nuclear crisis, and also leading to difficulties in Turkey's relations with Israel. The AKP's more critical view of Israel encouraged Turkey to become more vocal in its support for nuclear disarmament including in NPT diplomacy.[13] Departing from Turkey's past policy of abstention in deference to its friendly relations with Israel, in recent years Erdoğan aligned Turkey with Egypt's NPT rhetoric to press Israel to give up its nuclear weapons and join the NPT.[14] During the run-up to the May 2010 NPT Review Conference, Turkish President Abdullah Gül announced Turkey's support for a nuclear-weapon–free zone in the Middle East including Israel, and during the Review Conference Turkey joined Arab states in pressing for this outcome. Since then, Turkey has expressed strong support in NPT diplomacy for efforts to convene a conference in Helsinki toward that goal.

But while the AKP's Islamist outlook provides a rationale for greater Turkish commitment to disarmament, especially in the Middle East, Turkey has remained NATO-centered, which means that the reality of Turkey's security arrangements has deterred Ankara from going beyond disarmament rhetoric. Turkish officials involved in current NPT diplomacy state privately that as long as nuclear weapons exist in the world, Turkey's support for disarmament efforts must be qualified by Ankara's NATO commitments. A number of non-Turkish government officials and others involved in the 2015 NPT review process said that Ankara's credibility and influence in disarmament affairs will be minimal so long as Turkey retains 60–70 B61 theater nuclear weapons on its territory.

Turkish officials queried beginning in 2012 expressed the view that these weapons should not be removed without a consensus agreement among NATO members, and preferably under the condition that all such weapons in other NATO countries likewise be removed. Some Turkish officials asserted that Turkey would not insist on keeping the weapons if upon consultations NATO partners decided that removing them would best serve the alliance's interests. Still other Turkish experts opined that, particularly in view of the threats from unconventional weapons in the Middle East, the weapons systems in Turkey should be retained. One Turkish official referred to the B61s as "political weapons," meaning that, irrespective of their military utility, they serve the purpose of demonstrating NATO's defense commitment to Turkey.

The current situation was summed up in October 2013 by one senior Turkish official as follows:

> Turkey is committed to nuclear disarmament but we will follow NATO by choice. There should be a balance. We cannot dis-invent nuclear weapons. At the end of the disarmament process there might be one country with nuclear weapons. If everyone else is disarmed, that country can threaten the rest. So we are realistic about our ambitions in disarmament. Our NPT policy has to be seen in this context.

"NATO is the vehicle for our position on disarmament, including on nuclear-weapon–free zones," the official said. That statement has the ring of contradiction, and in the view of one senior NPT diplomat from a Western European country, it is a contradiction: "If Turkey's support for [nuclear-weapon–free zones] is anchored by a nuclear umbrella, it won't be credible," he said.

Turkey is a member of the Non-Proliferation and Disarmament Initiative (NPDI), a group of twelve states—seven of which are under the U.S. nuclear umbrella—formed after the 2010 NPT Review Conference to foster transparency in the nuclear-weapon policies of the nuclear-weapon states and to reduce the role of nuclear weapons in their defense strategies. The NPDI offered Turkey a forum to broadcast its support for nuclear-weapon–free zones, for the Comprehensive Nuclear Test Ban Treaty, and for the negotiation of a future Fissile Material Cut-Off Treaty. But the NPDI shares Turkey's own conflict between aspirations and realities. It isn't clear how much leverage and influence the NPDI will have. Its activities and statements might be largely brushed aside by the nuclear-weapon states.[15] A key aspiration of the NPDI is to secure acceptance by all nuclear-weapon states of a proposed "standard reporting form" on their nuclear-weapon postures and inventories. So far, it has not succeeded.

Turkey's NPT diplomacy during meetings since 2012 that were convened to prepare for the 2015 NPT Review Conference revealed the lack of correspondence between Ankara's disarmament aspirations and its embrace of NATO defense posture. "Nations should rely on…cooperation dialogue among themselves rather than the deterrent impact of nuclear arms," its 2012 statement concluded.[16] In 2013, Turkey's statement ended by asserting: "Global peace and security can only be achieved through common vision and interdependence, not so-called 'nuclear deterrence.'"[17]

NPT diplomacy in 2015 and beyond will be profoundly shaped by a more assertive Russia and a Middle East continuing to be ravaged by conflict. How Ankara interprets events unfolding in Russia and the Middle East will influence how Turkey deals with its disarmament/NATO conundrum. Since the 1990s Ankara and Moscow have worked to improve their post–Cold War bilateral relationship. In economics and especially energy commerce, the relationship has in fact been transformed, including on civil nuclear power, where the two states are now partners.[18] If Russia in the future follows the overall post–Cold War trajectory and recedes as a nuclear threat, Ankara will see more benefit in relinquishing the nuclear weapons on its territory and therefore have more space to join other NPT non-nuclear-weapon states in support of disarmament initiatives. If, however, Russia is seen as increasingly aggressive and prepared to use its nuclear arsenal as a means of intimidating others, Turkey will be further inhibited from going beyond rhetoric in support for disarmament in its NPT diplomacy. Turkey's relations with Russia became appreciably tense as the Ukraine crisis escalated in early 2014, and Ankara's reactions to these developments were aligned with those of its NATO partners.[19]

Likewise, if the Iran nuclear crisis is resolved peacefully, Turkey's security will increase, thereby enlarging its willingness to participate in nonproliferation diplomacy but perhaps without the urgency that since 2003 has driven NPT parties in the region, including Turkey, to press Israel to give up its nuclear weapons. Were negotiations with Iran to fail and instead Iran were to resume aggressive deployment of sensitive nuclear technology, Turkey's threat perception of a nuclear-armed Iran might increase, leading Ankara perhaps to edge closer to U.S. and NATO partners' version of events. But if an intensified Iran crisis were to escalate into a war, under AKP rule Turkey would likely deplore any Israeli or U.S. attack on Iran. A war and/or intensified nuclear proliferation in the Middle East might prompt some Turkish rhetorical expression of interest in having nuclear weapons. That interest might become operationalized should U.S. influence in Turkey and in the region dramatically decline, but currently the reality is that, barring unforeseen security challenges, Turkey now and for the indefinite future will have no inclination to develop nuclear weapons, and it will have very limited access to the technical means to independently do so.

Nonproliferation

In the years since Turkey ratified the NPT and endorsed on that occasion the halting of "proliferation of all kinds," Ankara has joined all multilateral arrangements established to combat the spread of weapons of mass destruction: the Comprehensive Nuclear Test Ban Treaty, Chemical Weapons Convention, Biological Weapons Convention, Nuclear Suppliers Group, and Missile Technology Control Regime. In addition, Turkey is a member of the Wassenaar Arrangement for export controls on conventional arms, it signed the global Arms Trade Treaty in July 2013, it actively participates in the U.S.-sponsored Proliferation Security Initiative, and it has attended all three nuclear security summits. Turkey has an NPT safeguards agreement with the IAEA and, beyond that, an Additional Protocol.

Throughout its NPT history, Turkey has supported the strengthening of multilateral nuclear export controls and the IAEA safeguards system. But there have been singular exceptions, principally informed by Turkey's bilateral relationship with Pakistan. More recently, in light of the AKP's foreign policy, Turkey's commitment to nonproliferation has been successfully challenged by Iran.

In its NPT diplomacy Turkey frequently expresses concern about and support for compliance with international binding nonproliferation and verification commitments. A senior Turkish official said in an interview that this concern "is based on our experience in our region, especially in the Middle East. A lot of proliferation-related activity has been going on there. That was true in the past and it has been true recently."

Turkey has especially been worried, this official said, about proliferation activities in two states with which Turkey shares a border: Iraq and Syria. It is no coincidence that these two states have also been at the heart of Turkey's concern about incursion from armed Kurdish insurgents. In the late 1990s war was threatened between Turkey and Syria over Syrian support for the Kurdistan Workers' Party (PKK), considered by Ankara to be a terrorist organization. A year before the Syria-Turkey crisis flared up, North Korea began assisting Syria's nuclear program.[20] During the previous two decades, the Soviet Union had provided Syria the means to produce chemical weapons and Damascus was attempting to make these.[21] The revelations beginning in 1991 that Iraq had an ambitious and secret nuclear-weapon program likewise prompted Turkish worry that unsecured Iraqi nuclear assets could fall into the hands of terrorists. At the same time, Russian, EU state, and Turkish intelligence agencies were discovering instances where nuclear and radioactive materials had been smuggled onto Turkish territory, in some cases with the participation of Turkish middlemen. When in 1994 it was confirmed that perpetrators had succeeded in smuggling small quantities of highly enriched uranium and plutonium from Russia to locations in Western Europe, Turkey's leaders openly expressed concern that a black market in stolen Soviet nuclear goods was emerging; according to one Western contemporary observer, Turkish officials claimed that Kurdish insurgents might obtain and use these materials.[22] Turkey, in partnership with NATO allies, other states, and the IAEA, has since undertaken more systematic and coordinated actions to respond to these concerns. Between 1994 and 2010, Turkish law enforcement agencies made 67 arrests in response to 104 recorded cases of illicit trafficking of nuclear and radiological materials.[23]

Shortly after the AKP came to power in 2002, the world's proliferation concern quickly shifted to Iran. The United States and Israel in that year informed the IAEA that Iran had secretly set up a uranium enrichment plant; the IAEA also learned that Iran had embarked on a project to build a reactor well-designed for the production of weapons-grade plutonium. Turkey urged Iran to comply with its IAEA safeguards obligations, and it supported diplomacy led by the United States and the European Union to resolve the Iran nuclear crisis peacefully. At the same time, Erdoğan tried to adjust Turkey's response to accommodate his Islamist party's interest in better relations with Iran by dismissing allegations of undeclared nuclear activities by Iran as "rumors." Behind these clumsy words were broader Turkish concerns that unless diplomacy succeeded, Israel and the United States might precipitate a war with Iran that could spread and threaten Turkey—just as Turkey had experienced in the wake of the deeply unpopular 2003 U.S.-led invasion of Iraq.

The AKP's foreign policy strongly factored in Turkey's response to the gradually escalating Iran nuclear crisis. Under Erdoğan and his foreign minister, Ahmet Davutoğlu (who succeeded him as prime minister in August 2014), beginning in 2002 the AKP pursued a vision of a Turkey at the center of a strategically crucial region and fully engaged with its neighbors to resolve all bilateral conflicts and grow their economies. Together with Brazil, which harbored compatible diplomatic aspirations, Turkey injected itself into the Iran nuclear crisis as a mediator. The effort culminated in the 2010 Tehran Declaration, which proposed that Iran would indefinitely remove a stockpile of low-enriched uranium from Iran to Turkey in exchange for an agreement by three Security Council powers and the IAEA to provide Iran fabricated fuel for an existing isotope-producing research reactor. The United States and other powers negotiating with Iran ultimately brushed this off and instead enacted sanctions against Iran in the UN Security Council over the objections of Turkey and Brazil.

In an interview, a senior Turkish official played down this event as a demonstration that Turkey and other middle powers had little influence over the conduct of big powers' nuclear diplomacy, but other Turkish sources reported that the AKP's leadership considered the failure of the Tehran Declaration to take hold as a bitter foreign policy defeat. The episode further strained relations between Ankara and Washington. Erdoğan thereafter reiterated AKP positions opposing the threat of force and economic sanctions against Iran, against the views of Security Council powers, which were convinced that sanctions and military threats were essential means to force Tehran to negotiate.

Response to Procurement by Pakistan and Iran

Lack of Turkish influence over the course of nuclear diplomacy with Iran did not prevent Turkey from being challenged by Iran's aggressive and clandestine nuclear behavior. In 2013 Germany requested Turkey's cooperation in prosecuting an Iranian nuclear smuggling ring. Turkey did not comply, seriously qualifying its commitment to nonproliferation and recalling past difficulties experienced by the U.S. government in encouraging Ankara to halt procurement for Pakistan's nuclear program.

According to Leon Fuerth, a former U.S. State Department official and member of the U.S. National Security Council, Pakistan's nuclear program began getting Turkish assistance in the late 1970s and a "particularly tenacious nuclear supply relationship" persisted "through the 1990s."[24] Another former U.S. official said in an interview that Washington had pressed Ankara for more than a decade to intervene to stop exports of nuclear dual-use items to Pakistan's uranium enrichment program that had been provided by two Turkish firms, EKA Elektronik Kontrol Aletleri and ETI Elektroteknik. This assistance to Pakistan's nuclear effort was finally addressed by Ankara by the end of the 1990s, but long after Turkey was first informed by the United States that termination of this trade was a high bilateral priority, and before the full extent of the Turkish firms' contribution to the black market network masterminded by Pakistani scientist A. Q. Khan was disclosed.[25]

In the background of Turkey's exports of nuclear dual-use goods to Pakistan looms a history of close bilateral ties. These were cultivated by the founder of the Turkish Republic, Mustafa Kemal Atatürk, and they extend to the present day.[26] Beyond that historical fact, narratives offered to explain Turkish-Pakistani nuclear commerce differ.

Fuerth suggests that the Turkish supply relationship with Pakistan tracks with difficulties and crises in the overall U.S.-Turkey bilateral relationship and with efforts of Turkish leaders to move away from the West.[27] He also argues that Turkey may have permitted nuclear cooperation with

Pakistan because of the desire of leaders to "balance" Turkey's foreign policy by becoming "less dependent" on the United States, and he concludes that it is "possible that the delay [in Turkish NPT ratification] was evidence of reservations or even a desire to keep the nuclear [weapon] option open."[28]

Turkish officials in interviews in 2012 and 2013 categorically denied that Turkey has ever considered a nuclear-weapon option and, further, asserted that Ankara's membership in NATO since 1952 obviated such considerations. During the 1980s and 1990s, when according to U.S. officials Turkish entities and individuals were engaged in dual-use nuclear trade with Pakistan, the Department of State raised concerns about this matter with Turkey based on intelligence findings from other U.S. government agencies. The intelligence, sources recalled in interviews, was considered highly reliable and included the transcripts of recorded conversations between representatives of the implicated Turkish firms and senior Turkish government officials. U.S. intelligence background reports and evaluations prepared for negotiators' guidance did not suggest that the Turkish-Pakistani nuclear trade was motivated or informed by any Turkish interest in nuclear weapons or by any official Turkish participation in Pakistan's nuclear-weapon–related activities. According to a former U.S. government official, it was assumed instead that "corruption relationships" involving Turkish political leaders and the company executives who had been fingered by U.S. intelligence—not secret Turkish nuclear-weapon ambitions—ultimately explained "why it took them so long to stop this relationship."[29] More generally, Sinan Ülgen has suggested that the absence of a high Turkish nonproliferation profile and lack of engagement by Turkish officials and government agencies at that time meant that until Turkey's U.S. relations were on the line, Ankara didn't prioritize Turkish companies' connections to Pakistan.[30]

But there were also problems in getting the U.S. government to take firm action. One former U.S. official said that during the 1980s, nonproliferation officials in the U.S. government believed the allegations against the Turkish perpetrators were serious enough to warrant halting aid to Turkey under terms of the U.S. Foreign Assistance Act, which barred recipients from assisting foreign nuclear-weapon programs. That step was never seriously considered at the top of the U.S. government, he said, because the

White House's priority in its relationship with Pakistan beginning in 1979 was to get its support to extract the Soviet Union from Afghanistan.[31]

Turkish media and politics have long speculated about the existence of a highly secretive, anti-democratic Turkish "deep state" running Turkey from behind the scenes. Modern Turkish leaders, including Erdoğan, have been cited by Turkish media as claiming that such secret ruling connections have existed and have functioned in Turkey as far back as the Ottoman Empire.[32] Some Western accounts have speculated that, until Erdoğan made far-reaching changes in how Turkey is governed, Turkey's civilian-military National Security Council, established in 1960, may have been the locus of undisclosed "deep state" activity.[33] Fuerth singles out the National Security Council for mention in his treatment of the question whether Turkey would consider reaching for nuclear weapons.[34] In private interviews, several researchers asserted that Turkey's recent political history includes activities by conspiratorial and clandestine cabals and alliances, and one asserted, without providing any details, that some participants advocated that Turkey acquire a nuclear-weapon capability.

But there is no historical evidence in the public domain of Turkish activity to establish such a capability, and the publicly available record likewise indicates that Turkey has little or no infrastructure to support nuclear-weapon development. Indeed, according to former officials knowledgeable about the IAEA's implementation of Turkey's NPT safeguards agreement, including its Additional Protocol, the IAEA does not have any evidence or any findings, made by its personnel or provided by IAEA member states, pointing to past or present undeclared Turkish nuclear activities. These sources acknowledge, however, that, prior to the mid-1990s, when discovery of Iraq's clandestine nuclear program prompted the IAEA to begin focusing in earnest on signs of countries' hidden nuclear work, the IAEA would not likely have detected undeclared activities at locations not specifically included on Turkey's inventory list as subject to safeguards.[35] But today, following from implementation of its Additional Protocol, Turkey has a "broader conclusion" from the IAEA, implying that, after the IAEA completed a holistic and historical examination of Turkey's entire nuclear program, it was confident that all nuclear activities in Turkey were declared, understood, and dedicated to peaceful uses.

The IAEA issued the broader conclusion for Turkey in 2012, eleven years after it began implementing Turkey's Additional Protocol, an unusually long period of time for a country without nuclear power infrastructure or any declared and sensitive fuel-cycle installations. The broader conclusion for Turkey was delayed for several reasons. These included a comprehensive IAEA investigation of Turkey's historical interest in many activities for the front end of the nuclear fuel cycle; the absence of an effective nuclear regulatory body during much of the country's nuclear history; and a probe of Turkish firms' participation through the Khan network in foreign nuclear programs including Iran and Libya, following revelations about these activities after Turkey's Additional Protocol entered into force in 2001. According to knowledgeable sources, the delay in reaching the broader conclusion was not due to allegations or discovery of undeclared or clandestine nuclear activities, including any activities related to the development of nuclear weapons.

While Turkey and the IAEA in the 2000s were probing Turkish firms' black market activities, Ankara's resolve to implement high nonproliferation standards was again challenged by Iran. The new AKP-led government sought to avoid bilateral friction with Iran over fresh allegations of undisclosed Iranian nuclear activities, but the IAEA and its member states, including Security Council powers, learned that Iranian nuclear buyers were shopping in Turkey.

In 2013 Germany requested the extradition of Hossein Tanideh, an Iranian citizen who had been arrested in Turkey after Iranian companies operating in Turkey were investigated by the Turkish authorities in cooperation with foreign, mostly Western, governments. Since 2012, Tanideh had been included on lists of individuals compiled by the United States and the European Union as subject to sanctions and asset seizure related to procurement activities for Iran's nuclear program. The German request was not honored by Turkey, and the matter set up a conflict between Ankara's nonproliferation commitments and other Turkish political aims.

Tanideh is implicated in German criminal proceedings against individuals who were convicted in November 2013 of having exported 91 nuclear-related items from Germany to Iran in violation of sanctions and export controls.[36] According to Turkish media reports, perpetrators provided false end-user statements claiming that goods destined for Iran would be shipped instead to other destinations including Turkey.[37] Tanideh was arrested in Turkey in January 2013 on the basis of a notice served by Interpol. A senior Turkish official interviewed in October 2013 said it would be up to the Turkish judiciary whether to honor Germany's extradition request.

According to Western officials in early 2014, Iran pressured Turkey to release Tanideh and allow him to return to Iran, and Turkish news reports asserted that Turkish intelligence and justice agencies had conferred over the German request but preferred instead to put Tanideh on trial in Turkey. In March 2014 a Turkish official confirmed information from other sources that a decision had been made by Turkey on Germany's extradition request and that the record of that decision had been withheld from the public domain.

Knowledgeable officials interviewed in May 2014 said that Turkey released Tanideh from custody and that it is assumed that Tanideh has returned to Iran. In September 2014, sources disclosed that Tanideh had been allowed to leave the country thanks to the intercession of Turkish government intelligence personnel who were cooperating with counterparts in Iran. Without providing details, these sources said that Tanideh's release was consistent with unconfirmed published reports asserting that the head of Turkey's national intelligence agency, Milli İstihbarat Teşkilatı, had passed on to Iran sensitive U.S. and Israeli intelligence data as part of an effort by Erdoğan to forge a closer bilateral working relationship between Turkey and Iran.[38]

According to Western officials, Tanideh is not an insignificant player in Iran's ongoing clandestine nuclear procurement effort. The evidence that Tanideh-controlled firms in Turkey were critical to efforts by Iran to smuggle NSG-listed goods from Germany to Iran "was considerable and made known to the Turkish government," one official said. "If Iran pressured Turkey to get him released, you can assume he was very valuable to Iran."

Whether Tanideh's release was personally approved by Erdoğan or not, if confirmed, the incident will be judged a setback for Turkey's nonproliferation profile, and it will suggest that Erdoğan sacrificed Turkey's commitment to nonproliferation on the altar of other interests, including Ankara's bilateral relationship with Tehran. Turkey's diplomatic conflict over Tanideh came to pass at a time when the Nuclear Suppliers Group is considering steps to reduce proliferation risk in brokered transactions involving third countries and in trade involving transshipments, and when Turkey is advocating a closer relationship between the Nuclear Suppliers Group and Pakistan.

As was the case for Turkey's foray into the big powers' Iran nuclear diplomacy a few years earlier, the AKP's foreign policy aspirations also were at play in Turkey's decisionmaking in the Tanideh case. Both episodes harmed Turkey's relations with Western governments. In 2010 Erdoğan's nuclear diplomacy reaped very public and rhetorical discord with these states over Iran sanctions. Four years later, Erdoğan again angered Western powers, this time by trying to balance irreconcilable elements of the AKP's policy toward Iran. At the working level, Turkish export control and nonproliferation officials are aware of the potential threat posed by Iran's unbridled nuclear development, including to the AKP's entire foreign policy project of prioritizing the regional economy over security issues.[39] On the other side of this complex ledger, especially under Iranian pressure such as that reported in the Tanideh case, the AKP may be tempted to cooperate with Iran to combat Kurdish separatists, limit damage over differences on Iraq and Syria, conclude lucrative trade deals, or compensate for other irritants such as Turkey's decision to deploy technology related to NATO missile defense.[40]

No such dilemmas have beset Turkish policymakers concerning Ankara's consistent support for IAEA verification. Turkey lined up quickly after the Gulf War in 1991 to support the IAEA in developing a more intrusive safeguards approach, which in 1997 became the Additional Protocol. In 1994, Turkey told the IAEA it strongly supported efforts to detect undeclared nuclear activities because Turkey is "still suffering from the economic and social losses incurred as a consequence" of Iraq's pre-1991 nuclear program.[41] Currently, with its broader conclusion on IAEA safeguards in place, Turkey supports efforts by the IAEA secretariat to universalize the Additional Protocol and to further strengthen IAEA safeguards through the elaboration and implementation of the so-called state-level concept. Turkey has not joined Argentina, Brazil, Russia, and some non-aligned states in raising objections to aspects of the IAEA's safeguards approach. Some states critical of the state-level concept warn that extending its scope will put the IAEA's independence at risk because the IAEA will become more reliant upon intelligence gathering by a handful of largely Western countries led by the United States. While aware that its nuclear intelligence-gathering infrastructure pales by comparison to assets wielded by the United States, Israel, and a few other states, Turkey does not support this view.

Peaceful Uses of Nuclear Energy and NSG Membership

From the very beginning of Turkey's NPT membership, Ankara has stressed that non-nuclear-weapon states, in exchange for their non-proliferation commitment, should be afforded nuclear cooperation on the basis of nondiscrimination. As Turkey's interest in developing nuclear power has grown, its defense of NPT parties' Article IV rights has become more elaborate and specific. In recent years this has led to conflicts between Turkey's policies and the nonproliferation goals of the United States and some other advanced nuclear NPT states—indeed, to internal conflicts among Turkey's own NPT policy goals.

Twenty years after joining the NPT, Turkey joined the Nuclear Suppliers Group, the world's leading multilateral nuclear export control mechanism. Three developments contributed to this event: rising expectations about the role nuclear power should play in Turkey's energy production, growing Turkish industrial capability to produce NSG-listed goods, and steps taken by Turkey to resolve U.S.-Turkish friction over Turkey's exports to Pakistan.

Like many other countries responding to America's post–World War II "Atoms for Peace" initiative, Turkey during the 1950s set up a national nuclear research organization and then obtained a U.S.-supplied nuclear research reactor. After Turkey signed the NPT and before ratifying it, Turkey's next move was to launch a nuclear power project on the basis of negotiations with foreign companies. Negotiations with several European vendor firms were under way when Turkey ratified the NPT.[42] On the eve of ratification, the U.S. firm General Dynamics provided Turkey's second research reactor, ITU-TRR.[43] A short time later, Turkey added a third unit, TR-2.

During the next two decades Turkey negotiated with firms in Canada, Germany, and the United States to obtain access to nuclear power technology. These efforts did not succeed. Turkey tried again in the 1990s, on the basis of an ambitious projection from the Turkish Atomic Energy Authority that the country should aim for an installed nuclear power generation capacity of 34,000 megawatts.[44] This effort likewise failed.

Some Turkish analysis and commentary blamed foreign government interference—especially from the United States—for Turkey's failure to conclude power reactor supply contracts during the 1980s and 1990s.[45] U.S. officials acknowledge that Washington's support for U.S. firms' efforts to sell nuclear equipment to Turkey was limited by the unresolved bilateral dispute over trade with Pakistan. But they and European counterparts said that the main reason for Turkey's failure was that all foreign vendors doubted Ankara's commitment to cover the political risk should Turkey's first nuclear power plant project be canceled before it was finished.[46] More recent Turkish analysis concludes that Turkey's nuclear power quest failed for commercial—not nonproliferation—reasons.[47]

The true picture is more nuanced. Specific transactions and planned cooperation relationships between Turkey and other states provoked objections from the United States, but Turkish plans to import power reactors from established vendor states were not resisted by these governments. During the 1980s, for example, Turkey and Argentina discussed several possible nuclear cooperation projects. The United States objected to plans for joint development and marketing of reactors, in part because of concern that Pakistan might benefit via its bilateral relationship with Turkey. But Turkish interest in importing a turnkey power reactor from Argentina was not seen as controversial because both the reactor and its fuel would be under IAEA safeguards. The United States at this time was far more concerned about Argentina's own nuclear aspirations, especially considering that country's rivalry with Brazil. Unlike Turkey, neither Argentina nor Brazil was a party to the NPT, and both were engaged in sensitive uranium enrichment and reprocessing research.[48]

Throughout the early history of Turkey's quest for nuclear power, Ankara had no nuclear cooperation agreement with the United States. Had Turkey selected a U.S. vendor to build power reactors, such an agreement would have expedited realization of the project. A green light from Washington for negotiation came only after Turkey addressed long-standing Pakistan-related nuclear trade issues at the end of the 1990s. By

2000 it was understood that Turkey would join the Nuclear Suppliers Group and negotiate a bilateral nuclear trade agreement with Washington.[49]

The trade agreement was concluded in 2002 but not approved by the U.S. Congress until 2008. Some Turkish analysts have asserted that residual Turkey-Pakistan issues held up approval by U.S. lawmakers, but a senior Turkish official said in September 2013 that by 2004 both countries were on the same page concerning the A. Q. Khan network,[50] and he concurred with a Turkish analysis that said that objections raised by Turkey to U.S. efforts beginning in 2004 to limit nuclear newcomer states' access to enrichment and reprocessing technology accounted for the delay.[51] A U.S. official specified that initially, Turkey also had been critical of terms in the draft bilateral agreement that afforded the United States prior consent rights over enrichment and reprocessing activities involving U.S.-obligated nuclear material—a feature of most U.S. bilateral nuclear cooperation agreements.[52]

Turkey had been a member of the Nuclear Suppliers Group for four years when President George W. Bush announced, in the aftermath of the revelations of the Khan network and the 9/11 terrorist attacks, that the United States sought to ban the spread of enrichment and reprocessing activities by countries that until then had not deployed those technologies. Turkey then joined all other NSG members in opposing the U.S. initiative. In 2008, the United States abandoned this effort in favor of a compromise to establish new NSG criteria for the export of enrichment and reprocessing items by states adhering to NSG guidelines.

The ensuing discussion inside the NSG set up a conflict for the first time between two NPT principles—nonproliferation and access to nuclear technology based on nondiscrimination—that had long guided Turkish NPT diplomacy and policies. The United States, supported by some others in the group, aimed to establish a rule that any new enrichment plant outside of the small group of countries that currently exploit this technology commercially should be based on a "black box" principle, meaning that an advanced-country technology holder may lease equipment to, but not share know-how with, the country where a new plant is located. A senior Turkish official said in an interview that Ankara opposed making the "black box" model obligatory; Turkey was joined, he said, by Argentina, Brazil, Canada, South Africa, and Switzerland.

Like many other states that believe the peaceful use of nuclear energy is a sovereign right, Turkey based its policy on this issue "on principle, not because Turkey wanted to enrich uranium," the senior Turkish official said. According to other nuclear trade diplomats, some Turkish politicians and government officials suggested in media interviews that Turkey should include enrichment in a suite of long-term nuclear industry options, justified by the Turkish Atomic Energy Authority's forecast that Turkey would build more than two dozen power reactors. AKP leaders were also sympathetic to efforts by Ankara to resist foreign pressure on Turkey to limit its nuclear technology options, along the lines argued by Iran in its conflict with the United States and other Western states since 2003, stressing that all NPT parties had the unrestricted right to peacefully exploit nuclear technology.[53] This had been in fact Turkey's view since 1980. Only in 2011 was the conflict in the Nuclear Suppliers Group over this issue resolved by consensus of all 46 NSG members. Turkey's position served as the common denominator; the "black box" approach was recommended by new NSG guidelines but not made obligatory.

During the same decade, Turkey and the United States fought in the NSG through another conflict: Washington's advocacy beginning in 2005 for an exception to the NSG guidelines for India. Turkey had joined the NSG in 2000 after Ankara and Washington had worked toward settling their differences over nuclear trade with Pakistan. Facing the U.S. quest for a nuclear "India deal," Ankara's close relations with Islamabad once again were a potential bone of contention. After three years of internal discussion, in which the United States and India applied diplomatic pressure to recalcitrant NSG participants, including Turkey, the NSG in 2008 approved the exception for India by consensus. Prior to that decision, Turkey had expressed reservations about uniquely exempting India from the NSG guidelines.

Two Nuclear Suppliers Group participants with strong bilateral ties to Pakistan—China and Turkey—did not block consensus on this issue in 2008. Since then, China has challenged the credibility of the NSG by expanding the scope of bilateral nuclear trade with Pakistan under what Beijing says is a nuclear trade understanding reached with Pakistan before China joined the NSG in 2004. Turkey has refrained from engaging in any nuclear-related commerce with Pakistan, but it now strongly advocates within the group a closer relationship between Pakistan and the NSG in the interest of getting Islamabad to adhere to the NSG guidelines and, politically, balancing India's access to global nuclear markets with a similar accommodation for Pakistan.[54] Ankara contends that improving Pakistan's relationship with the NSG would allow the country to adopt a more conciliatory position at the Conference on Disarmament, thus giving a boost to multilateral nuclear disarmament initiatives.

Turkey in recent years also opposed the idea, advocated by the IAEA, the United States, and some other advanced states in response to Iran's nuclear development, of setting up multilateral fuel banks to dissuade countries from enriching uranium. In Turkey's view, these could serve as an instrument to limit states' right of access to enrichment and reprocessing.[55] Turkey objected that the proposed fuel banks would be a disincentive for states to enrich uranium; advocates at IAEA and NPT conclaves pointed out that discouraging enrichment was in fact what the fuel banks were intended to do. A senior Turkish official said that Turkish sensitivity about U.S. advocacy of efforts to limit enrichment and reprocessing may have been heightened by the historical memory of a U.S. arms embargo imposed after Turkey invaded Cyprus in 1974.[56]

Turkey's positions on Iran since the beginning of the crisis over that country's nuclear program have exemplified Ankara's effort to balance recent conflicts that emerged between the nonproliferation and peaceful use pillars in its NPT policy. The Iran crisis began one year after the AKP took power. Turkey adopted a position in between Iran and the West, by upholding Iran's Article IV rights, while supporting the United States and other advanced and Western states in criticizing Iran's failure to comply with its NPT safeguards agreement. Turkey likewise sought a middle ground on sanctions imposed on Iran, in principle allowing for multilateral penalties imposed by the UN Security Council but opposing any unilateral sanctions enacted by the United States and other Western powers and, in 2010, voting against multilateral sanctions in the Security Council in a specific instance. Ankara adopted the view that Iran's nuclear program did not directly threaten Turkey but that Turkey supported international efforts to resolve the Iran crisis peacefully, toward the result that Iran would enjoy its Article IV rights and also meet its nonproliferation obligations under Articles I and III.

CONCLUSION

When Turkey signed the NPT in 1969, the potential conflicts inherent in the treaty's aspirations were the concern of relatively few. That was no longer true by the time Turkey ratified the treaty in 1980. The same year, the escalating Cold War prompted an acrimonious debate over disarmament at an NPT Review Conference, and for the first time, the parties failed to successfully negotiate a final declaration. Since then, Turkey's NATO membership has stood in the way of Ankara's lifting its disarmament profile in its NPT diplomacy. Unlike some other NATO members, Turkey is not under pressure domestically to play a greater disarmament role, and given lingering uncertainty that conflicts in the Middle East may escalate into wars and that the threat posed to Turkey by Russia may reemerge, Turkey's disarmament aspiration appears indefinitely held hostage by the unwillingness of the nuclear-weapon states to get rid of their nuclear arms. Regardless of the desire of Turkish leaders to use the NPT process as a lever to encourage the powers to disarm, as long as Turkey relies on NATO to guarantee its security, this contradiction will bedevil the making of Turkish NPT policy.

Like many other states, Turkey during the 1960s was encouraged by its U.S. ally to join the NPT. Like others, Turkey benefited from U.S. nuclear cooperation, even before Ankara ratified the treaty. When it signed the treaty, Turkey had no nuclear technology to proliferate and hence no nonproliferation conflict. As Turkey developed as an industrial state, problems in its relationship with the United States eventually spilled over into Ankara's NPT diplomacy. Irritation with the United States may have dissuaded Ankara from quickly ratifying the treaty for several years. Friction with Washington over Turkey's ties with Pakistan inhibited the United States from actively supporting U.S. industry to advance Turkey's quest for nuclear power, but most of Turkey's difficulties in realizing its nuclear power ambitions were infrastructural, financial, and commercial in nature. In 2000 the United States was the key to rebalancing Turkey's nonproliferation and peaceful-use aspirations: After Ankara addressed U.S. concerns about Turkish firms' nuclear trade links to Pakistan, the United States and Turkey concluded a bilateral nuclear trade agreement and the United States blessed Turkey's entry into the NSG—a step making more likely that in the future Ankara would firmly control its exporting industry. Then came 9/11 and the A. Q. Khan revelations. U.S. reaction to these events led to conflict with Turkey inside the NSG, as U.S. policy to further restrict nuclear activities was viewed by Ankara as a challenge to Article IV.

Turkey's growing interest in nuclear power, friction with the United States, and frustration with repeated failures to negotiate with foreign firms on what it held to be favorable terms, lined up Ankara alongside developing and nonaligned countries that by the 2000s were asserting their rights to peaceful-use nuclear activities. This posed few direct problems for Ankara's relationship with the United States since U.S. nuclear technology was no longer seen by Turkey to be essential to its development. Instead Turkey partnered with Russian firms.

Turkey's stance during most of the Iran crisis since 2003 appeared closer to that of Russia than to that of the United States. Guided by the AKP, Ankara, like Moscow, openly advocated Iran's Article IV rights, including its right to enrich uranium; favored a negotiated peaceful solution to the crisis; and opposed unilateral U.S. sanctions. In other nonproliferation areas, however—fuel banks and IAEA safeguards—Turkey and Russia have remained apart.

If Turkey remains beset by the contradiction between its NATO security interests and its commitment to disarmament, Turkey finally resolved the conflict between its nonproliferation aspirations and its commitment to close ties with Pakistan, though doing so required two decades. The United States facilitated this at the end of the 1990s by opening the way to Turkish participation in the Nuclear Suppliers Group and by negotiating a bilateral nuclear trade agreement. Since then, Turkey has been broadly successful in balancing its support for both nonproliferation and peaceful nuclear cooperation, but on specific occasions Turkey has demoted its nonproliferation commitments in favor of bilateral relationships and other political considerations.

Iran today holds the key to the nuclear future of the region of which the AKP asserts Turkey is at the center. So far, and inspired since 2002 by the AKP's foreign policy agenda, Turkey has scrupulously avoided confrontation with Iran, but that has come at the price of Ankara's failing to accommodate the nonproliferation expectations of key NATO and bilateral partners. Turkey was not legally obligated by its NPT or NSG commitments to prosecute Iranian nuclear smugglers operating on its territory. The Tanideh case underlines that, as in its past dealings with Pakistan, Ankara may take decisions that protect specific bilateral relationships at the expense of its nonproliferation profile.

The AKP may govern Turkey for a long time to come, and its influence on policy should not be underestimated. Turkey has a foreign policy with an Islamic pedigree designed to maximize Turkey's operational independence especially by focusing on developments in its region, by emphasizing trade over security, and that is wary of big powers' interventions in its neighborhood. Including for these reasons, as Henri Barkey says, under the AKP Turkey has a "more benign view of Iran," regardless of significant differences on specific issues. The two countries' bilateral relationship is a balancing act with numerous interests at play, and "by and large, the Turkish-Iranian relationship will soldier on."[57] Turkey may well conclude that cooperating with Iran in the interest of combating Kurdish separatists, vying for influence in Iraq, forging greater trade ties, and creating a bigger space between itself and Washington, are together more important than addressing Iran's nuclear threat. Turkey also may be willing to risk displeasing its Western partners in conducting its relations with Iran if it concludes that Western states would ultimately rather see Turkey than Iran "win in Iraq."[58]

In the big picture, regardless of the AKP's economic bias, Turkey's security perception largely informs its NPT policies. That sets up conflicts between its aspiration for disarmament and its place under NATO's umbrella. The same security concerns prevent Turkey from lining up programmatically with the Non-Aligned Movement on important issues. Ankara's position on Article IV and Iran seems to match the Non-Aligned Movement's own position, but Turkey's firm support for the IAEA's verification role under NPT Article III—especially its endorsement of the Additional Protocol as a component of the current global nonproliferation standard—departs from the Non-Aligned Movement narrative and is in fact identical with the position held by Turkey's NATO partners.[59]

NOTES

1   Erik J. Zuercher, *Turkey: A Modern History* (London: I. B. Tauris, 2004), 276.

2   Turkish Atomic Energy Authority, "TR-2 Research Reactor Operation Unit," October 4, 2010, www.taek.gov.tr/en/institutional/affiliates/cekmece-nuclear-research-and-training-center/310-nuclear-facility-division/869-reactor-operation-unit.html.

3   Leon Fuerth, "Turkey: Nuclear Choices Amongst Dangerous Neighbors," in *The Nuclear Tipping Point: Why States Reconsider Their Nuclear Choices*, ed. Kurt M. Campbell, Robert J. Einhorn, and Mitchell B. Reiss (Washington, D.C.: Brookings Institution, 2004), 149–50; Mustafa Kibaroğlu, "Turkey," in *Europe and Nuclear Disarmament: Debates and Political Attitudes in 16 European Countries*, ed. Harold Mueller (Brussels: European Interuniversity Press, 1998), 179.

4   Nicole Pope and Hugh Pope, *Turkey Unveiled: A History of Modern Turkey* (London: Overlook Duckworth, 1998), 101–104.

5   Ibid., 124.

6   United Nations Office for Disarmament Affairs, "Turkey: Ratification of Treaty on the Non-Proliferation of Nuclear Weapons (NPT)," April 17, 1980, http://disarmament.un.org/treaties/a/npt/turkey/rat/london.

7   U.S. State Department memorandum in the Lyndon Baines Johnson Presidential Library, "Views of Various Countries Consulted About the Proposed Nuclear Nonproliferation Treaty," February 12, 1967.

8   Duygu B. Sezer, "Turkey's New Security Environment, Nuclear Weapons and Proliferation," *Comparative Strategy* 14, issue 2 (1995): 156.

9   Ibid., 150–51.

10 Elliot Hentov, "Turkey and Iran," *Turkey's Global Strategy*, London School of Economics IDEAS Special Report, May 2011, 30, www.lse.ac.uk/IDEAS/publications/reports/pdf/SR007/iran.pdf. In a memorable parliamentary vote on March 1, 2003, Turkish lawmakers denied the United States access to Iraq from Turkey. While some experts reject the conclusion of Ian Kearns (see Ian Kearns, "Turkey, NATO and Nuclear Weapons," Occasional Paper, Royal United Services Institute and European Leadership Network, January 2013, 5, www.rusi.org/downloads/assets/OP_201201_Turkey,_NATO_and_Nuclear_Weapons.pdf) that the vote led to the "worst crisis in U.S.-Turkish relations in decades," in view of assistance provided by the United States to Turkey in anticipation that Turkey would incur costs from the Iraq War, "others identified it as the beginning of a policy of distancing Ankara from U.S. influence in the region" (see Hentov, "Turkey and Iran," 30). The Justice and Development Party (AKP), which took power with a large parliamentary majority after the 2002 election, may have permitted representatives to vote as they wished regardless of the government's formal position in support of providing the U.S. access to Iraq from Turkish territory. According to some contemporary observers in interviews, Turkish military officials, in a campaign they launched to discredit the AKP, encouraged lawmakers not to give the United States the access it desired.

11 Sinan Ülgen, "Turkey and the Bomb," Carnegie Paper, Carnegie Endowment for International Peace, February 2012, 12, http://carnegieendowment.org/2012/02/15/turkey-and-bomb#.

12 Kearns, "Turkey, NATO and Nuclear Weapons," 10.

13 The AKP's distancing from Israel also prompted Turkey to resent Israel's December 2007 bombing of a site that Israel and the United States identified as a clandestine nuclear reactor in Syria. Israel reportedly apologized to Turkey for having violated Turkish airspace during the mission to attack the target in Syria. See Amos Harel, "Five Years On, New Details Emerge About Israeli Strike on Syrian Reactor," *Haaretz*, September 10, 2012, www.haaretz.com/news/diplomacy-defense/five-years-on-new-details-emerge-about-israeli-strike-on-syrian-reactor-1.464033. At the time of this event, Turkey still supported the Baathist regime in Syria. Turkish officials were briefed by U.S. and Israeli counterparts after the fact that Syria's nuclear project had benefited from assistance from North Korea. A Turkish official said that the incident posed a dilemma for Ankara since Turkey opposed both the Israeli incursion as well as the apparent proliferation event, and that Turkey shared the view of the IAEA's director general that the attack on the installation in Syria represented a vote of no confidence by Israel in IAEA safeguards.

14 Sinan Ülgen, "Preventing the Proliferation of Weapons of Mass Destruction: What Role for Turkey?" Center for Economics and Foreign Policy Studies, Discussion Paper Series 2010/2, June 2010, 10.

15 George Perkovich, "Reducing the Role of Nuclear Weapons: What Can the NPDI Do?" Carnegie Endowment for International Peace, November 27, 2012, http://carnegieendowment.org/2012/11/27/reducing-role-of-nuclear-weapons-what-npdi-can-do.

16 "Statement by Turkey at the First PrepCom of the 2015 NPT RevCon," Vienna, April 30–May 11, 2012, www.un.org/disarmament/WMD/Nuclear/NPT2015/PrepCom2012/statements/20120430/PM/Turkey.pdf.

17 "Statement by Turkey at the Second Prepcom for the 2015 NPT Review Conference," Geneva, April 22–May 3, 2013.

18 Kearns, "Turkey, NATO and Nuclear Weapons," 7.

19 Adam Balcer, "Dances With the Bear: Turkey and Russia After Crimea," Working Paper 8, Center for East European Studies, University of Warsaw, July 2014, www.iai.it/pdf/GTE/GTE_WP_08.pdf. This analysis suggests that Russo-Turkish relations may be troubled by two parallel developments: Turkey's criticism of Russian behavior vis-à-vis Ukraine, and an anticipated reduction in coming years of Turkey's dependence on Russia for supply of energy fuels.

20 Private communication from United States government official, June 2013.

21 Mary Beth D. Nikitin, et al., "Syria's Chemical Weapons: Issues for Congress," Congressional Research Service, September 30, 2013, www.fas.org/sgp/crs/nuke/R42848.pdf.

22 Private communication from Western government customs intelligence official, December 2010.

23 "Turkey, Balkans Fight Nuclear Trafficking," *Turkish Weekly*, January 3, 2014, www.turkishweekly.net/news/160975/turkey-balkans-fight-nuclear-trafficking.html.

24 Fuerth, "Turkey: Nuclear Choices Amongst Dangerous Neighbors," 160–64. Fuerth's account, however, also suggests that U.S. President Ronald Reagan succeeded in halting Turkish firms' assistance to Pakistan's uranium enrichment program after Reagan prioritized this issue with Turkey in 1988. A former U.S. government official in September 2013 said that Turkey did not put an end to the black market assistance

to Pakistan's program until a decade after Reagan met with Turkish President Kenan Evren in 1988. "Turkey failed to cooperate with the U.S. over this the whole time Reagan was in office," he said.

25 Former U.S. government official, private communication, September 2013.

26 Mustafa Kibaroğlu, "Turkey's Quest for Peaceful Nuclear Power," *Nonproliferation Review* 4, no. 3 (Spring–Summer 1997): 35.

27 Fuerth, "Turkey: Nuclear Choices Amongst Dangerous Neighbors."

28 Ibid., 150.

29 Former U.S. government officials, private communications, July 2012 and November 2013.

30 Ülgen, "Preventing the Proliferation of Weapons of Mass Destruction: What Role for Turkey?" 15.

31 Former U.S. government official, private communication, December 2013.

32 Yusuf Kanli, "The Turkish Deep State," *Hürriyet Daily News*, January 29, 2007, http://arama.hurriyet.com.tr/arsivnews.aspx?id=-598627.

33 Pope and Pope, *Turkey Unveiled: A History of Modern Turkey*, 310ff.

34 Fuerth, "Turkey: Nuclear Choices Amongst Dangerous Neighbors," 165.

35 Former verification officials, private communications, March, October, and December 2013.

36 "Covert Iranian Nuclear Dealings via Turkey Revealed," *Today's Zaman*, March 12, 2013, www.todayszaman.com/news-309539-covert-iranian-nuclear-dealings-via-turkey-revealed.html.

37 "Haftstrafen für Kaufleute wegen illegaler Iran-Geschäfte" [Prison Sentences for Merchants Over Illegal Deals With Iran], *Hamburger Abendblatt*, November 9, 2013, www.abendblatt.de/politik/deutschland/article121709622/Haftstrafen-fuer-Kaufleute-wegen-illegaler-Iran-Geschaefte.html.

38 Adam Entous and Joe Parkinson, "Turkey's Spymaster Plots Own Course on Syria: Hakan Fidan Takes Independent Tack in Wake of Arab Spring," *Wall Street Journal*, October 10, 2013, http://online.wsj.com/news/articles/SB10001424052702303643304579107373585228330.

39 Hentov, "Turkey and Iran," 30.

40 Henri Barkey, "Turkish–Iranian Competition After the Arab Spring," *Survival* 54, no. 6 (2012).

41 Report by the Director General [of the 1995 IAEA General Conference] on Strengthening the Effectiveness and Improving the Efficiency of the Safeguards System; GC(39)/17, 28.

42 Kibaroğlu, "Turkey's Quest for Peaceful Nuclear Power," 35.

43 "Research Reactor Details— ITU-TRR, TECH UNIV," www-naweb.iaea.org/napc/physics/research_reactors/database/RR%20Data%20Base/datasets/report/Turkey,%20Republic%20of%20%20Research%20Reactor%20Details%20-%20ITU-TRR,%20TECH%20UNIV.htm.

44 Kibaroğlu, "Turkey's Quest for Peaceful Nuclear Power," 38.

45 Ibid., 34ff. See also Kibaroğlu, "Turkey and Shared Responsibilities," in *Shared Responsibilities for Nuclear Disarmament: A Global Debate*, ed. Scott D.

Sagan (Cambridge, Mass.: American Academy of Arts and Sciences, 2010), www.mustafakibaroglu.com/sitebuildercontent/sitebuilderfiles/kibaroglu-daedalus-sharingresponsibilities-revised-edited-final-12march2010.pdf. "Turkey's plans for building nuclear power plants have been obstructed by its Western allies, fearful that Turkey would one day decide to weaponize its capabilities.… These fears stem from rumors regarding Turkey's close relations with Pakistan.…" (36).

46   Mark Hibbs, "Turkey Pulls Back on Reactor, Europe's Vendors Skeptical," *Nucleonics Week* (May 23, 1996): 9.

47   Sebnem Udam, "Turkey's Nuclear Comeback," *Nonproliferation Review* (June 16, 2010): 366.

48   "Turkey—Power Balance Concerns," in *Nuclear Programmes in the Middle East: In the Shadow of Iran* (London: International Institute of Strategic Studies, 2008), 61.

49   According to a former U.S. government official in September 2012, U.S. negotiators of Turkey's bilateral nuclear cooperation pact were not aware of any quid pro quo between Turkey's resolution of the Pakistan nuclear issue and the decision by the NSG to include Turkey as a participant. Another U.S. official involved in the bilateral nuclear trade negotiations said, however, that "it can be assumed that if Turkey wanted to join the NSG and conclude a [nuclear cooperation agreement] with the U.S., we wouldn't have looked favorably on either step so long as the issue [concerning trade with Pakistan] wasn't resolved." While by 2000 the NSG was prepared to welcome Turkey as a member, and the United States was getting ready to negotiate with Ankara the terms of a bilateral nuclear trade agreement, at that time the full extent of Turkish firms' participation in the Khan network, including their outsourcing of equipment production to locations outside Turkey, had not been revealed; that required three more years of criminal justice and intelligence investigations, including interviews with suspected Turkish businessmen after Khan's arrest in 2003. Kibaroğlu, "Turkey's Quest for Peaceful Nuclear Power," 44, mentions possible Greek objections to Turkish NSG participation in advance of Turkey's joining the group.

50   Interview with a senior Turkish official, September 2013.

51   Sinan Ülgen, ed., *The Turkish Model for Transition to Nuclear Power* (Istanbul: Center for Economics and Foreign Policy Studies, 2011), 151–56, www.edam.org.tr/edamnukleer/edamreport.pdf.

52   Private communication from U.S. official, April 2013.

53   Private communication from U.S. diplomat, June 2007.

54   Interviews with Turkish nuclear trade and nuclear policy officials, April 2012 and September 2013.

55   Ülgen, "Turkey and the Bomb."

56   Interview with senior Turkish official, September 2013.

57   Henri Barkey, "Turkey's Perspectives on Nuclear Weapons and Disarmament," in *Unblocking the Road to Zero*, ed. Barry Blechman (Washington, D.C.: Stimson Center, 2009).

58   Barkey, "Turkish–Iranian Competition After the Arab Spring," 149.

59   A senior Turkish official said in an interview in October 2013 that "The NAM continually urges us to vote with them [in IAEA diplomacy] but we repeatedly explain to them that we don't share all their key interests."

CHAPTER 7

TURKEY AND NUCLEAR WEAPONS
Can This Be Real?

MUSTAFA KIBAROĞLU

INTRODUCTION

Since the dramatic revelations of Iran's illicit nuclear activities in 2002, commentators have speculated that Turkey will follow suit and seek to acquire nuclear-weapon capabilities to balance Iran and meet the potential challenges of a proliferation cascade in the Middle East. Academics and pundits have scrutinized Turkey's interest in nuclear energy projects with a view to "discovering the real motives" behind its past and current initiatives. What is the likelihood of Turkey "going nuclear" in the years ahead despite its outstanding performance under the nuclear nonproliferation regime?

A number of factors are believed to have kept Turkey from seeking to produce its own nuclear-weapon capabilities. A discussion about Turkey's domestic interests and characteristics as well as international factors, such as its membership in the North Atlantic Treaty Organization, its adherence to the nuclear nonproliferation regime, and its European Union (EU) vocation, will be followed by scenarios about possible courses of action that Turkish policymakers might adopt in case they decided to acquire nuclear weapons. Needless to say, this will be a speculative, intellectual exercise, which will be carried out with a view to addressing a number of issues that are most frequently raised inside and outside of Turkey, such as who would be in charge of a Turkish nuclear-weapon development project, what would be the strategy for evading Turkey's commitments under the nuclear non-proliferation regime, which capabilities and technologies would be acquired and/or indigenously developed for becoming self-sufficient in the long term, who would be the international partners, and the like.

THEORIES OF NUCLEAR PROLIFERATION AND THE CASE OF TURKEY

Nine countries in the world are acknowledged as possessing nuclear weapons. Seven of them—the United States, Russia, United Kingdom, France, China, India, and Pakistan—have proven their capabilities by carrying out nuclear tests. North Korea, meanwhile, also carried out tests, but its weaponization capacity is not known for sure. There is also Israel, which has neither acknowledged nor denied the existence of nuclear weapons in its arsenal.

Considering that approximately 200 states exist in the world, the number of states that are believed to possess nuclear weapons constitutes indeed a small fraction, less than 5 percent. The number of states, however, could have been much higher, perhaps in the dozens. Preventing a significant number of states from pursuing nuclear weapons would have been very difficult, if not impossible, had there been no international efforts that culminated in the Treaty on the Non-Proliferation of Nuclear Weapons (NPT).[1] The incentives as well as assurances that were given to states to join the NPT and to forgo the option of developing nuclear weapons in return for ample support in advancing peaceful applications of nuclear energy were found to be satisfactory by many states that might have otherwise pursued nuclear weapons.[2] Still, some states are suspected of acquiring scientific and technological capabilities that are necessary for developing nuclear weapons. This raises an essential question as to why states would want nuclear weapons.

The dynamics of nuclear proliferation have been explored by many academics.[3] In this sense, the curiosity about the motivation that leads countries to pursue nuclear-weapon programs is well researched.[4] First and foremost, it was found that countries attempt to do so for security reasons, to seek the capability to deter military threats. Neorealist Kenneth Waltz argued that military leaders dislike uncertainty and so claim that they need nuclear power to guard against an uncertain future. Many of the academics focus on external factors such as perceived threats. They maintain that it is a rational response for countries to protect their own interests for state survival. Classical realists, such as Zachary S. Davis and Richard K. Betts, contend that states pursue nuclear-weapon programs if doing so contributes to their own national security, but they accept that domestic politics also play a role in states' decisions. Thereby, they agree that there are different types of states that react differently to external factors.

Classical realists, unlike neorealists, posit that states have multiple, interlinked goals and that these objectives have both domestic and international aspects. Domestic concerns may vary from political stability to social cohesion, or economic strength to technological development. Politicians try not only to survive in the international arena but also to stay in power as long as possible. Thus, they may attempt to use nuclear-weapon programs as a tool to stay in power by winning public support and enhancing their domestic political position. Governments facing oppositional challenges might use the nuclear card to divert public criticism. In this sense, nuclear weapons can be utilized as an instrument to mobilize a nation's patriotism.[5]

Additionally, states may decide to acquire nuclear weapons in a quest for regional and/or global hegemony. According to the realist assumption, states try to maximize power and therefore undertake such programs because of their desire to achieve regional preponderance. The realist camp argues that these states may even attempt to blackmail other states into submission to their political wishes as a nuclear power. They may want to acquire nuclear weapons to gain prestige and also in order to join the club of "untouchables," and they believe that their nation can attain complete independence only by becoming a nuclear-weapon state.[6]

Moreover, states may do so in order to strengthen their economies, to try to get "more bang for the buck."[7] In other words, they compare conventional weapons in terms of the security output per dollar. It is here that the belief comes into play that states possessing nuclear weapons invest less on conventional weapons. However, Harald Müller and other scholars argue that nuclear-weapon states still invest in conventional weapons and that acquiring nuclear arms is not necessarily accompanied by decreasing investments in conventional arms.[8]

Furthermore, the bureaucracy, namely the military and scientific establishments, may push the governments in power to embark on nuclear-weapon programs. Persuading decisionmakers to pursue such a policy may get more resources for the bureaucracy, which also seeks to enhance its position vis-à-vis other national institutions.

Finally, the high-tech environment attributable to advances in science and technology enables such programs to be carried through. When the know-how is present, it is easier for bureaucratic institutions to persuade the government to go nuclear.

Against this background, which only briefly explains how and why states are motivated to pursue nuclear weapons, one may argue that these factors have been, in varying degrees, present in the case of Turkey as well. For instance, Turkish security elites have always claimed that Turkey is situated in a very dangerous neighborhood, being at the epicenter of the Balkans, the Caucasus, and the Middle East. That's how they justified maintaining a large standing army, ranking second among the North Atlantic Treaty Organization (NATO) countries after the United States, in order to be able to cope with threats perceived from all directions. As a result, the defense budget in Turkey has constituted a significant proportion of government spending for decades.[9] Under these circumstances, Turkish decisionmakers could have justified developing nuclear weapons as both a deterrent to enemies and a way of diminishing military expenditures.

Since the mid-1980s, starting with the government of President Turgut Özal, Turkish politicians have had the ambition to elevate Turkey to the position of a regional power and a global player in world politics. From that perspective, the prestige that would be gained from nuclear-weapon capability would have made them more confident in their pursuit of assertive foreign and security policies to achieve their objective.

Moreover, the scientific and scholarly community as well as the civilian and military bureaucracy in Turkey have been generally receptive to the idea of acquiring "nuclear power," even if "nuclear weapons" may not have been explicitly pronounced in that context. Many scholars and experts as well as bureaucrats and politicians in Turkey have promoted plans for developing a scientifically and technologically advanced nuclear infrastructure and a complete nuclear fuel cycle apart from installing nuclear power reactors for energy generation.[10]

THE FACTORS THAT KEPT TURKEY FROM GOING NUCLEAR

The foregoing brief profile suggests that Turkish political and security elites could have found justification for pursuing a nuclear weapon capability. However, they did not do this for a number of reasons. Turkey's international considerations, such as its membership in NATO, its adherence to the nuclear nonproliferation regime, and its EU vocation would have made it difficult for Turkish decisionmakers to pursue nuclear weapons even if they had wished to do so.

Turkey has long pursued a policy of subscribing to the relevant international arms control and disarmament treaties and conventions as well as contributing to their effective implementation. This practice helped raise a cadre of civil and military bureaucrats, scholars, scientists, experts, and intellectuals who have developed a stance against the proliferation of weapons of mass destruction. Diplomats, military officers, and bureaucrats from the Ministry of Foreign Affairs, Ministry of National Defense, Ministry of Energy and Natural Resources, the General Staff, and the like, who have been involved in international nonproliferation efforts in various platforms, have over the years become highly conscious of the possible consequences of developing nuclear weapons clandestinely. Therefore, since the 1970s, whenever plans for building nuclear power plants have come to the fore, the Turkish bureaucracy endured a certain degree of tension with respect to a number of critical decisions, such as what should be the type and/or the size of the nuclear reactors, which country should be the supplier, whether or not to invest in uranium enrichment and plutonium reprocessing in the future, and the like.

For instance, Turkey and Argentina signed an agreement in 1990 to form a joint architectural-engineering firm to develop Argentina's modular low-power CAREM-25 reactors, one in each country.[11] While Turkey agreed to provide most of the financing for the work, Argentina would provide most of the technology.[12] If preparations for the building of two units had gone ahead as planned, work on the first unit in Argentina would have begun in 1991, and the construction of the second unit, in Turkey, would have begun in 1992.[13] The long-term goal of the joint venture was to export the reactor to other nations in Latin America, Africa, and the Middle East.[14]

Turgut Özal, as Turkish prime minister prior to becoming president, had corresponded and met with Argentine President Carlos Menem regarding the project and thus played a key role in obtaining the agreement. Despite the fact that high-level talks in the nuclear field had been carried out between Argentina and Turkey and had culminated in a formal document, the CAREM-25 project was canceled a year later by the unilateral decision of Yalcin Sanalan, then director of the Turkish Atomic Energy Authority. Sanalan notes that he "found the prospects of the CAREM-25 deal ambiguous" on the grounds that "CAREM-25 was too small for electricity generation and too big for research or training, however, very suitable for plutonium production" and thus a proliferation concern. Therefore, Sanalan "concluded that such an ambiguous project would decrease the chances of Turkey in its current and future quest for large-scale nuclear power plants which the country really needed."[15] This anecdote, shared with the author by Professor Sanalan, may provide insights into the inner circles of the decisionmaking mechanism in the Turkish state bureaucracy with respect to the differences of opinion on sensitive issues, such as nuclear energy projects, that are still relevant today.[16]

A major impediment to the potential ambitions of Turkish decisionmakers to acquire nuclear-weapon capabilities has been the security assurances given by NATO to Turkey since its entry into the alliance in February 1952. Article 5 of the North Atlantic Treaty signed on April 4, 1949, in Washington, D.C., constitutes the basis of the "positive security guarantees" given to Turkey by other members of NATO, that an attack on any member is an attack against all. Accordingly, Turkey's entire territory has been covered by a nuclear umbrella that may effectively deter possible attacks from other countries. U.S. nuclear weapons that have been deployed in allied countries in Europe including Turkey have long been at the crux of the "extended deterrence" capability of the alliance.[17] The

NATO-wide ballistic missile defense system, known as the "missile shield," is another form of assurance provided to Turkey by the alliance against the threat of proliferation of weapons of mass destruction and their delivery vehicles. This system is expected to become fully operational in 2018. The decision to install the essential parts of the missile shield (the radar site in Kürecik in the Malatya district in eastern Turkey) was finally made at the NATO summit meetings in Lisbon in November 2010. The radar site in Kürecik started to operate as a NATO asset concomitantly with the Chicago summit meeting of NATO in May 2012.[18]

A second factor that limited Turkey's options has been its treaty obligations under the nuclear nonproliferation regime. Turkey became a state party to the NPT by signing it on January 29, 1969, and subsequently ratifying it on April 17, 1980. Turkey also concluded a "full-scope" Safeguards Agreement with the International Atomic Energy Agency (IAEA) in 1982, meaning that the agency monitors all nuclear facilities in Turkey. Eventually, Turkey joined the other international nuclear nonproliferation efforts such as the Zangger Committee and the Nuclear Suppliers Group (NSG) in 2000, and it signed and ratified the Comprehensive Nuclear Test Ban Treaty in 2001. Turkey also endorsed efforts to strengthen the nuclear nonproliferation regime and the verification mechanism of the IAEA.

Accordingly, in the 1990s, Turkey paid much attention to the proceedings of a study called "Programme 93+2" as an attempt to make IAEA safeguards inspections more intrusive, which culminated in the Additional Protocol in 1998. Turkey signed the Additional Protocol on June 6, 2000, and it entered into force for Turkey on July 17, 2001.[19] Turkey also cosponsored a joint working paper by a number of European allies and submitted it to the May 2010 NPT Review Conference. The paper stated that global nuclear disarmament requires an incremental but sustained approach in which all treaty-based nuclear arms control and disarmament agreements are indispensable for the active promotion of collective security and cooperation in the pursuit of this objective.[20] More recently, Turkey took part in the Non-Proliferation and Disarmament Initiative (NPDI), a cross-regional, ministerial-level group initiated by Australia and Japan that focuses on practical steps with a view to taking forward the consensus outcomes of the 2010 NPT Review Conference.[21]

A third factor that limited the options available to Turkish decision-makers has been Turkey's candidate status before the European Union. Turkey began a "Westernization" process as early as during the nineteenth-century Ottoman Empire. The early modernization attempts in the Turkish Republic were undertaken during the 1920s with a view toward accelerating the country's transformation into a Western-style, secular nation-state.[22] Turkey's interest in its relations with the West gained further momentum with its admission to the Council of Europe in 1949 and to NATO in 1952. Then came the Ankara Agreement in 1963, which gave Turkey an attractive prospect to eventually become a full member of the European Union. Since 1987, when the first official application for membership was made, Turkey has been a consistent candidate for EU accession. After long deliberations, the accession negotiations between Turkey and the EU started in 2005 but soon stagnated. If and when the accession process is successfully completed, as a condition of full membership, Turkey would become a state party to the Euratom Treaty, which would permit only peaceful applications of nuclear technology.

U.S. NUCLEAR WEAPONS IN EUROPE AND THE POSITION OF TURKEY

Among the key factors that kept Turkey from considering the acquisition of nuclear weapons, the impact of NATO's "extended deterrence" and the role of the U.S. nuclear weapons deployed on Turkish territory in this context deserve special attention. As of fall 2014, the state of affairs between Turkey and its Western allies is testing solidarity in a number of issue areas—including the remaining U.S. tactical nuclear weapons stationed in Europe, where Turkey and its allies may have contradicting policies.

Reports indicate that 150–200 tactical nuclear weapons belonging to the United States are still deployed in five European members of NATO: Belgium, Germany, Italy, the Netherlands, and Turkey.[23] Turkey has hosted U.S. nuclear weapons since the intermediate-range Jupiter missiles were deployed in 1961 as a result of decisions made at NATO's 1957 Paris summit. The Jupiter missiles were withdrawn in 1963 in the aftermath of the Cuban Missile Crisis. After that, U.S. nuclear weapons under U.S. Air Force custody remained in air bases in Eskişehir, Malatya, Ankara, and

Balıkesir, where F-100, F-104, and F-4 Phantom aircraft belonging to the Turkish Air Force were planned to deliver them.[24] With the end of the Cold War, most of the nuclear weapons were withdrawn from these bases. Today, U.S. tactical nuclear weapons are still stored in Turkey, albeit in much smaller numbers and in only one location, the İncirlik base near Adana on the eastern Mediterranean coast of Turkey.[25]

Turkey has long relied on the strong deterrent capability of NATO's nuclear strategy. Turkish officials would prefer to continue to benefit from the extended deterrence provided by these weapons stationed in Turkey. However, the positions of some European allies have not been fully compatible with that of Turkey. On February 26, 2010, the foreign ministers of Belgium, Germany, Luxembourg, the Netherlands, and Norway wrote a letter to Anders Fogh Rasmussen, NATO's secretary general at the time, indicating that they would "welcome the initiative taken by President Obama to strive toward substantial reductions in strategic armaments, and to move toward reducing the role of nuclear weapons and seek peace and security in a world without nuclear weapons." The letter also emphasized that there should be discussions in NATO as to what the allies "can do to move closer to this overall political objective."[26]

In response, Turkish officials warned that such an attitude would seriously damage "solidarity" and "burden sharing," two fundamental principles of the alliance that have been the basis for Turkey's agreeing to the deployment of U.S. nuclear weapons on its soil. Turkish officials, therefore, expected other allies also to continue hosting U.S. nuclear weapons on their soil, if only in symbolic numbers. In this way, Turkey would not stand out as the only NATO country in Europe that retains U.S. nuclear weapons.[27]

Notwithstanding the initiative of the five European allies, the Strategic Concept Document that was adopted during the Lisbon summit in November 2010 underlined that NATO would remain a nuclear alliance as long as nuclear weapons exist in the world.[28] Moreover, in the Deterrence and Defense Posture Review issued at the Chicago summit in May 2012, the allies agreed that the North Atlantic Council would task the appropriate committees to develop concepts for how to ensure the broadest possible participation of allies concerned in their nuclear sharing arrangements, including in case NATO were to decide to reduce its reliance on nuclear weapons based in Europe.[29]

Voices are still being heard in Europe suggesting that these weapons should be sent back to the United States despite the adopted document that emphasized the significance of the forward-deployed U.S. nuclear weapons on European territory. For instance, in advance of national (and European) elections in Belgium in May 2014, the Flemish socialists announced at a party congress that they would not enter a Belgian government if U.S. tactical nuclear weapons remained on Belgian territory, let alone were modernized.[30]

Russia's invasion and annexation of Crimea probably will end the debate within NATO over withdrawing U.S. nuclear weapons. However, if in the future the European allies decide to ask the United States to withdraw its nuclear weapons from their territory, Turkey would be left with two options: to carry the burden of forward deployment of U.S. nuclear weapons within the alliance all by itself (or perhaps with Italy) and to reverse its long-standing policy of hosting U.S. nuclear weapons and send them back to the United States. Each option is worth analyzing.

If the U.S. tactical nuclear weapons were withdrawn from other European allies, Turkish authorities, despite their declaratory policy, could welcome the continuation of deployment of these weapons in Turkey as a credible deterrent to current and potential rivals in the region. It goes without saying that the proliferation of weapons of mass destruction in the Middle East, in particular, constitutes one of the primary sources of threats to Turkey's national security and stability. Turkish officials are not fully confident that the assurances given to Turkey by NATO would be credible enough without the actual presence of the United States in the picture. Therefore, they may prefer relying on the extended deterrence provided by the United States and its nuclear weapons, whose deployment on Turkish soil would be seen as a guarantee of U.S. involvement on Turkey's side in any future contingencies.[31]

If Turkey decided to join European allies should they opt to send back the U.S. nuclear weapons, Turkish-American relations could be negatively affected. Nearly a decade ago, when asked about the status of U.S. nuclear weapons, Turkish officials underlined that keeping the weapons had to do with the nature of Turkish-U.S. relations and Turkey's place in the Western alliance. The deployment of tactical nuclear weapons that remained in Turkey was believed to strengthen the bonds between the United States and

Turkey, which had suffered serious setbacks due to the unfulfilled expectations on each side during and after the crisis situation in Iraq starting in late 2002.[32] Turkish-American bonds were severely strained after Turkey's rejection of the U.S. request to station troops on its soil in the run-up to the war in Iraq in March 2003.[33] Turkish officials then feared that withdrawing U.S. nuclear weapons from Turkey in the aftermath of such a delicate period would have further weakened the long-standing strategic alliance.[34] Hence, at a time when the two countries may need each other's tangible support in protecting their national interests vis-à-vis the current conflict-laden issues, such as Syria's civil war, Iran's controversial nuclear program, Iraq's instability, and Ukraine's ongoing crisis, sending U.S. nuclear weapons back may not hold much appeal for Turkish officials.

LONG-TERM RELIABILITY OF THE FACTORS THAT KEPT TURKEY FROM ACQUIRING NUCLEAR WEAPONS

Notwithstanding the long-standing responsible state practice of Turkish governments, which joined every existing nuclear nonproliferation instrument, concerns remain about whether the next generation of decision-makers in Turkey would consider taking steps toward acquiring nuclear weapons. These concerns may be prompted, among other factors, by the stance of the Justice and Development Party (AKP) government vis-à-vis regional and global issues that exhibit notable differences in style and substance from the policies of former governments. Especially over the past five years, Turkey has followed a much more assertive foreign policy with an ambition to become a more visible player on many regional and even global issues.[35] Ahmet Davutoğlu, the former foreign minister who became prime minister in August 2014, considers this policy as a necessary but not sufficient condition for the realization of his government's long-term objective "to make Turkey a global power."[36] Western security analysts are, therefore, concerned that Turkey's state practice and the current state of affairs in its relations with the institutions mentioned above, which are seen as insurance policies against its potential inclination toward acquiring nuclear capabilities, may not remain on the same track in the medium to long term.

With respect to the assurances provided by NATO, in the eyes of many Turks, the powerful image of the Alliance has been diluted in its transformation from a collective defense organization with a "hard power" stance, to a collective security organization with a perceived "soft power" attitude. Likewise, under the influence of anti-American sentiment, which is pervasive in the Turkish public domain lately, NATO is starting to be seen as an organization that "serves primarily the interests of the United States and helping it to establish its world hegemony."[37] Traces of such an approach can be seen in the harsh criticisms leveled against the missile shield project of the alliance that required the deployment of a radar site in Turkey.[38]

As for the European vocation, it is necessary to underline that even though the start of the accession talks has institutionally brought Turkey closer to the EU, the optimistic mood among Turks and Europeans soon took a negative turn. Suspicions of Turkey's suitability for membership have grown ever since.[39] Objections to Turkey's membership on the basis of identity-related considerations have increased, while the arguments in favor of Turkish accession on the basis of cost-benefit calculations have weakened.[40] With growing societal security concerns over the existence of millions of Muslims in Europe, the EU has become increasingly reluctant to develop a strong geopolitical commitment to Turkey's eventual accession.[41] Besides, the AKP government, which was commended in the West for being progressive as well as promoting the rule of law, improving human rights, normalizing troubled civil-military relations, boosting the economy, and taking genuine initiatives to solve Turkey's long-standing conflicts with its neighbors, is now being harshly criticized in Europe. For instance, during a visit to Brussels on January 22, 2014, Recep Tayyip Erdoğan, the prime minister at the time, was reminded by European Commission President José Manuel Barroso that "respect for rule of law and independence of the judiciary [were] basic principles of democracy and essential conditions for EU membership." Similarly, Herman Van Rompuy, then president of the European Council, told Erdoğan that "it is important not to backtrack on achievements and to ensure that the judiciary is able to function without discrimination or preference."[42]

Within the nuclear nonproliferation regime, meanwhile, a series of significant international developments over the past decade have cast doubts on the future prospects of the regime. These developments include North Korea's nuclear detonations; revelations about Iran's secret facilities suitable for fissile material production; failure to persuade all states of concern, including Iran, to ratify the IAEA's Additional Protocol; and failure to urge the enforcement of the Comprehensive Nuclear Test Ban Treaty. Hence, the possibility of the NPT's becoming an ineffective treaty stands out as one particular reason that some in Turkey espouse the idea of having at least the basic infrastructure for nuclear-weapon capability.[43] Under these circumstances, while Turkey, like other states, certainly could import nuclear power plants or seek over time to indigenously produce them without intending also to acquire nuclear-weapon capabilities, suspicions often arise that dual purposes lie beyond new nuclear power programs, especially those that might include uranium enrichment and plutonium reprocessing.

SCENARIOS ABOUT TURKEY'S NUCLEAR FUTURE

So far, in this chapter, issues concerning Turkey's official engagement with nuclear energy projects have been discussed, by and large, by relying on verifiable information, either collected by the author while conducting academic research on the subject over the past two decades, or on readily available open sources, such as academic books and journals as well as magazines and media pieces. Therefore, topics that are covered in this chapter until now have presented the author's interpretation of various features of Turkey's official stance toward the proliferation of nuclear weapons. Special attention has been paid to staying away from speculative comments with respect to the strategies that could have been adopted by Turkish governments.

This section, however, will attempt to discuss a number of scenarios[44] concerning the possible courses of action that Turkish policymakers might consider adopting in case they decided to pursue nuclear weapons in response to a number of contingencies. Such contingencies include Iran's manifesting its nuclear-weapon capability, which would embolden the clergy in their attitude toward Turkey; aggravation of the security situation in Iraq and Syria, with both holding Turkey responsible for such an outcome; worsening of relations with Russia due to its mounting pressure on Turkey to prevent the passage of American naval vessels through the Turkish Straits on the grounds of the provisions of the 1936 Montreux Convention;[45] deterioration of relations with the European Union due to European leaders' severe criticism of undemocratic practices of Turkish governments; and waning of the U.S. commitment to the strategy of extended nuclear deterrence. This list of contingencies can be further extended, as the geopolitical and geostrategic environment in Turkey's neighborhood is highly conducive to unwelcome developments, as most recently seen in Crimea.

In the event of one or more of these contingencies taking place, the scenario exercised here would assume that the Turkish ruling elite might decide to develop a nuclear-weapon capability as a countermeasure against the threats posed to Turkey's territorial integrity and political sovereignty in the medium to long term. Should this be the case, several critical questions would then emerge: Who would be in charge of the nuclear-weapon development project? What would be the strategy for evading Turkey's commitments under the nuclear nonproliferation regime? Which capabilities and technologies would have to be acquired and/or indigenously developed for becoming self-sufficient in the long term? Who could be the international partners?

Answers to these and other hypothetical questions may be given only by the authorities of the institutions that would be involved in such a project. Hence, the following paragraphs can only speculate about what the possible course of action might be.

Who Would Be in Charge of the Nuclear-Weapon Development Project?

It would be safe to argue that any decision to pursue nuclear weapons would be made first by consensus at the highest echelons of the state mechanism. Then, the National Security Council meetings, chaired by the president and attended by the prime minister as well as a few key ministers, such as foreign and defense ministers and their top bureaucrats, would most likely be the exclusive venue for high-level deliberations on alternative strategies. These meetings would also be appropriate for inviting a limited number of scholars, scientists, and experts whose opinions would be expected to help the top executives to crystallize their decisions about whether to stay in the NPT while working on a weapons program; which technologies to acquire and/or develop indigenously; which country or countries to collaborate with in the international arena, and so on. Of course, inviting such people might compromise secrecy. But the law forbids revealing information about meetings taking place at the National Security Council.

By virtue of the role that it used to play until the democratization of civil-military relations in Turkey, the National Security Council also would act as an interim general secretariat, responsible for coordinating the agenda of the meetings and keeping their records, at least until the ultimate strategic decision would be made as to who would be in charge of the project. This last point would be crucial, and the answer would depend on who is president and/or prime minister at the time the project would be under consideration. Turkey is a parliamentary democracy, and despite being the "commander in chief" according to the Turkish constitution, the president was usually considered as de facto subordinate to the prime minister when it comes to critical executive decisions (an exception being if the president is a charismatic leader like the late Turgut Özal or the current President Recep Tayyip Erdoğan).[46] According to the constitutional amendments of 2010, Turkey's president will henceforth be elected by popular vote, as was the case on August 10, 2014. This may change the power relationship between the president and the prime minister and may give an enhanced sense of policy initiative to new presidents. The prime ministry might then act only as the executive branch carrying out decisions that would be supervised by the president. Alternatively, the prime minister could lead the project. Charisma gains currency in this equation because it would be crucial both to motivate the top bureaucracy to perform its best as well as to achieve harmony within the state mechanism where diverging views and opposition to the project would be likely.

Not everyone who would be involved in these deliberations would be like-minded and fully support the idea to develop nuclear weapons. Opposition would most certainly come, primarily from the career diplomats in the Ministry of Foreign Affairs, who would inform the top decisionmakers about the possible serious consequences of Turkey's violation of its treaty obligations under the NPT, not to mention the degree of damage that such a move might cause to political, economic, and military relations with the Western alliance in the first place.

The General Staff, meanwhile, would try not to project an image of being an adamant supporter of the nuclear-weapon program. This would be done as a cautionary attitude, considering a possible backlash in due course in case the politicians reversed their decision, either under the constraints that they would face in the international arena, or due to a change in the government and/or political leadership, which may want to cancel the project altogether. The General Staff also would be cautious about the sensitive nature of military relations with NATO in general and the United States in particular. Yet, the degree of opposition, if only sporadic, that could come from the top brass might not be decisive enough to change the course of action if the country proceeded down the road toward nuclear weapons.

The National Intelligence Authority, even though it is represented in the National Security Council meetings at the undersecretariat level, recently has started to play a much more important role in the foreign and security domains, just as in the case of the Central Intelligence Agency of the United States and the Secret Intelligence Service (MI6) of the United Kingdom, due to the restructuring of its departments as well as its mission and its vision. The National Intelligence Authority would probably be in a standby position at all times with a view to establishing contacts for the procurement of the material and technology necessary for the advancement of the project as well as to provide, among others, counterespionage service to protect the scientific and technical staff involved in the project.

In addition to these institutions that would constitute the backbone, or the "A-Team," of the project, select ministries, such as the Ministry of Energy and Natural Resources, and government institutions, such as the Turkish Atomic Energy Authority, would be the tip of the iceberg—officially responsible for the execution of the nuclear project in the open in accordance with Turkey's treaty obligations, considering that Turkey would stay in the NPT at least until it was within three months of being able to produce a bomb, when it could give notice to withdraw.

What Would Be the Strategies for Acquiring the Necessary Capabilities?

For any country whose decisionmakers would consider developing an elaborate nuclear program that would enable them to divert peaceful applications to military applications in the future, two different strategies may be contemplated. The first would be to acquire the necessary technological infrastructure and scientific skills through legal transactions from supplier countries by staying in the NPT and then to walk away from the treaty obligations after having acquired the advanced capabilities necessary for a self-sufficient nuclear-weapon development program. The second would be to collaborate with one of the major nuclear-weapon-capable states, possibly one that would not be bound by NPT obligations and also willing to provide the technological infrastructure as well as the scientific skills necessary to build nuclear weapons clandestinely.

In the case of Turkey, policymakers might consider that a feasible option would be a combination of the two alternatives, which would mean to acquire, on the one hand, some necessary capabilities through legal transactions from the supplier states, as well as to acquire, on the other hand, scientific skills and technological parts that are necessary to build facilities for uranium enrichment and/or plutonium reprocessing, which would probably be denied to Turkey by its allies in the West. Recent developments taking place in the context of the Nuclear Suppliers Group (NSG) meetings only enhance this view.

For instance, when the United States proposed limiting transfers of enrichment and reprocessing technologies from the members of the NSG to states that do not already possess this technology, the NSG drafted a "clean text" in 2008 that attempted to take into account all viewpoints and finally receive the support of all members. The clean text included both objective and subjective criteria. The objective criteria were mandatory conditions that suppliers would have to take into account before completing an enrichment and reprocessing transfer. The subjective criteria were additional criteria that suppliers could take into account, such as those proposed earlier by the United States regarding considerations of domestic and regional stability; prior agreement to refrain from acquiring enrichment and reprocessing capability; a coherent reason for desiring the technology; and whether a transfer would be used for peaceful purposes. In addition to the opposition voiced by Brazil and Argentina, Turkey opposed the subjective criteria of the clean text. In particular, Turkey took issue with whether a plausible reason exists for a transfer of sensitive nuclear technology to take place, and the impact of the transfer on the country and the region's stability and security.[47] Turkey felt that it would be viewed as being in an unstable region, and therefore it would be denied transfers regardless of its nonproliferation record and commitments.[48]

Turkey's position was that the NSG should not victimize any country simply because its neighbors are considered problematic.[49] Turkey also opposed the "black-box" requirement for trade of sensitive technology,[50] which means that the transfer must take place under conditions that will not permit or enable replication of the technology.[51] As outlined in a recent Centre for Economics and Foreign Policy Studies report,[52] although Turkey has invested in a number of technologies needed to form the basis of a civilian nuclear energy program, its lack of commercial scale enrichment and reprocessing technologies make it unlikely that Turkey could quickly develop a nuclear weapon. The report also states that given the nascent state of its nuclear industry, as well as the difficulties involved with the development of commercial-scale enrichment and reprocessing, Turkey would likely have to rely on foreign suppliers for fuel-cycle technology.

Would Turkey Imagine Putting Weapons on Missiles?

Within the context of the scenario outlined here, there is also need for a discussion about whether Turkey would want to put weapons on missiles, and if so, which ones would be feasible.[53] At present, Turkey's missile capability is limited and consists of short-range rockets and missiles. Turkey had initially sought to partner with the United States for the co-development of a system similar to the U.S. Army's Tactical Missile System (ATACMS), but the two sides could never reach an agreement on the terms of technology transfer. Turkey purchased 72 MGM-140A ATACMS surface-to-surface missile batteries from Lockheed Martin in 1996. The missile has a 300 kilometer range and is outfitted with a Global Positioning System–aided inertial guidance system. With the assistance of technology transfer from Chinese companies, Turkey first produced the 100-km range T-300 Kasırga (Hurricane) artillery rocket. Turkey and China also cooperated on the development of Roketsan's J series short-range ballistic missiles. The J-600 T Yıldırım (Thunder) is based on China's WS-1 unguided rocket system. The missile is reported to have a range of 150 kilometers with a warhead weighing 480 kilograms. It is reported that the Turkish Armed Forces have six batteries in operation.[54]

Turkey is keen on taking a longer-term approach to its missile production programs and aims to develop the capability to manufacture the missiles locally. Therefore, Turkey is likely to continue in its efforts to secure favorable technology transfer agreements from foreign suppliers, rather than simply opt to purchase a slew of missile systems without receiving any co-production or transferred technology in return. In December 2011, then prime minister Erdoğan reportedly called upon Turkey's defense industry to develop the capability to produce long-range missiles. Comparing Turkey's arsenal to Iran's, Erdoğan noted that Iran has missiles with a range of over 2,500 kilometers, while Turkey's missiles are limited to 150 kilometers.[55]

The Turkish Armed Forces have reportedly begun working on the nation's first project to develop an intercontinental ballistic missile. A decision to launch the project was made in a meeting of the Defense Industry Executive Committee, headed by Erdoğan and General Necdet Özel, chief of the General Staff. The committee decided to form a satellite launch center that would have a twofold effect on Turkey's aerospace and military endeavors. First, the center will enable Turkey to place its own satellites in orbit, and second, the center will allow the Turkish military to launch missiles that can navigate outside of the Earth's atmosphere. Attaining an intercontinental ballistic missile launch capability is reportedly the chief aim of the satellite launch center. The Turkish Defense Ministry, the Undersecretariat for Defense Industries, and the Scientific and Technological Research Council of Turkey (TÜBİTAK) have been jointly working on the project for some time. The report said Turkey could cooperate with an undisclosed Eastern European country to develop the satellite launch center.[56]

Who Could Be the International Partners?

In a scenario where Turkish decisionmakers are presumed to opt for a nuclear-weapon program, it is clear that Turkey would need international partners to advance its scientific and technological capabilities also in enrichment, reprocessing, weapons design, metallurgy, and other areas from the level where they are now to a level where developing nuclear weapons would be within its reach. In the scientific domain, Turkey had indeed a large number of nuclear engineers and technicians who earned their degrees in Turkey as well as abroad, from the 1970s onward when Turkey launched its first plans to build nuclear power reactors, none of which, mainly due to political problems, was ultimately realized.[57] Authorities argue that there are enough nuclear engineers and technicians, though most of them are scattered around the world due to the lack of job opportunities in Turkey—enough, anyway, for running at least two nuclear reactors and the related facilities. Added to these, just as Iran did in its Bushehr power plant deal with Russia, Turkey started to send approximately 150 students to Russian institutions to pursue graduate degrees in the nuclear field. The figure is expected to reach 600, including undergraduate students.

In addition to running the reactors, skilled personnel will be needed for doing design and construction of all the components for nuclear weapons. Yet, the formation of a cadre of scientists who would be responsible for the execution of the project may not be the biggest problem. Acquisition of sensitive technologies, such as uranium enrichment, would necessitate outside support, and the most likely candidate, as has been the case with Iran's nuclear program, would be Pakistan.

Before counting on any support that might come from Pakistan, it must be noted that Islamabad has apparently committed not to proliferate anymore, especially since the A. Q. Khan incident.[58] Pakistani scientists and their connections are watched very closely in this domain. Yet, it must also be taken into consideration that Turkey and Pakistan have historically developed fairly strong relations. For almost any Turkish citizen, Pakistan used to be the one and only country truly friendly to Turkey. Revolutionary thoughts and the principles of Mustafa Kemal Atatürk, the founder and the first president of the Republic of Turkey, inspired the Pakistani people in their fight for independence against the British. As an indication of the countries' ties, Atatürk's name has been bestowed on a variety of institutes, libraries, and the most beautiful districts in Pakistan's big cities. Turkey's historic relations with Pakistan gained momentum with the 1964 Regional Cooperation for Development agreement that brought together Turkey, Iran, and Pakistan. The warm relations were further intensified in many respects in the aftermath of the military coup in Turkey on September 12, 1980. The military leaders of Turkey and Pakistan, President General Kenan Evren and President General Zia ul-Haq, respectively, paid a series of visits to each other's country until the latter was killed in a plane crash on August 17, 1988. These ceremonial visits increased the magnitude of sympathy and of trade and cooperation in many fields, including the civilian and military spheres, providing fertile soil for rumors to grow. Hence, when NATO blocked Pakistan's enrichment program in the early 1980s, President Zia ul-Haq reportedly opened talks with Turkey, taking advantage of his "brotherhood" with his Turkish counterpart, Kenan Evren. At the same time, Greek Prime Minister Andreas Papandreou reportedly said that "Pakistan expected Turkey to act as a transshipper of material for a nuclear bomb and would reciprocate by proudly sharing the nuclear bomb technology with Turkey."[59] At present, both Pakistan and Turkey are disturbed by the fact that Iran has advanced its nuclear capabilities and is increasingly closer to building nuclear weapons. This may well be a reason for Turkey to expect Pakistan, which has mastered its nuclear-weapon development capabilities, especially uranium enrichment technology, to satisfy its expectations for technology transfer in the nuclear field.

Brazil could be another candidate with which Turkey could collaborate in the area of acquisition of sensitive technologies. The two countries joined forces in May 2010 to negotiate the "Tehran Declaration" that aimed at lending transparency to Iran's controversial nuclear program. Brazil's significance, from Turkey's perspective, also stems from its history of seeking nuclear weapons in the 1970s, when it was entered in a stiff race with its neighbor and rival Argentina. As a result of the concomitant transition to democracies from dictatorships, the two countries have signed the Quadripartite Agreement with IAEA and joined the NPT, which freed them from their ambitions to build nuclear weapons.[60] Yet Brazil maintains its powerful stance against the provisions of the Additional Protocol of the IAEA. Considering that Brazil is a member of the NPT and also the NSG, a collaboration between Turkey and Brazil could be similar to the one in which Russia supplies Iran with sensitive technologies within the context of the rights of non-nuclear-weapon states envisaged in Article IV of the NPT. The Brazilian Navy, which started a nuclear propulsion program in the early 1980s, has developed enrichment technology using centrifuges. Even though most enrichment of the fuel fabricated in Brazil for its nuclear reactors is undertaken by Urenco in Europe or in the United States, enrichment at the Aramar Experimental Center in Iperó (São Paolo state), which remains a naval facility, continues, and it is reported to be at a 5 percent U-235 level.[61] So Brazil might provide a basis for Turkey to acquire enrichment technology that may be eventually developed indigenously by Turkish engineers and technicians.

Japan may also be a candidate for supplying Turkey with sensitive technologies. In early 2014, there were conflicting reports about possible collaboration between Japan and Turkey in the area of uranium enrichment and plutonium reprocessing. During Prime Minister Erdoğan's meeting with his Japanese counterpart, Shinzo Abe, a $22 billion deal was signed in Tokyo on the nuclear plant project, planned to be built in Inceburun on the Black Sea coast by a joint venture involving Japan's Mitsubishi Heavy Industries. The Japanese daily *Asahi Shimbun* reported on January 8, 2014, that a senior Japanese Foreign Ministry official claimed that "upon Turkey's demand a clause, which allows Turkey to enrich uranium and extract plutonium, [was] added in the nuclear agreement signed by the two nations."[62] Soon after, however, Turkey ruled out the prospect of enriching uranium as part of its nuclear program. Officials said "the government of Prime Minister Erdoğan [would] not develop uranium enrichment capability" and that "Turkey did not reach agreement with Japan to produce nuclear fuel." On the same matter, Turkish Energy Minister Taner Yıldız said that "Turkey [did] not have any project regarding nuclear fuel and uranium enrichment." In a briefing on January 8, 2014, he acknowledged that "Turkey sought to learn to produce nuclear fuel for its four planned reactors." But, he said, "Ankara's interest [did] not extend to establishing a uranium enrichment sector."[63]

CONCLUSION

Based on this scenario, which is presented as a speculative intellectual exercise, even if one considers for a moment that Turkey decided to develop nuclear weapons and also managed to get the support of a nuclear power, or that it successfully established a clandestine nuclear-weapon procurement network, what would be the role of nuclear weapons in Turkey's security and foreign policies? Would nuclear weapons enhance Turkey's security? Or would they simply hurt Turkey's interests?

Any attempt to illegally pursue, let alone acquire, a nuclear-weapon capability would be extremely damaging to Turkey's vital interests. Turkey is passing through a difficult domestic and international political conjuncture in which any number of sensitive issues (social, economic, political) may be carefully exploited by its rivals. Even if one considers for a moment that Turkey managed to acquire a nuclear-weapon capability, under which scenarios and against whom would these weapons have added value in Turkey's foreign and security policies? It is difficult to give a meaningful answer to this question.

Of Turkey's immediate neighbors, Iraq and Syria are both deeply immersed in internal conflicts, and the future of these regimes is bleak. Neither is likely to pose a threat that Turkey could not deal with by non-nuclear means. As for Iran, even if its nuclear-weapon capability someday upsets the balance of its relations with Turkey, that alone may not be justification for Turkey's countering with nuclear weaponry and going through all possible hardships to get there. And in any case, a nuclear-weapon-capable Iran would most likely be dealt with collectively by the rest of the international community, the United States and Israel being at the forefront.[64] Greece and Armenia are other potential countries with which Turkey had, and may have, problems in its foreign relations. While Greece and Turkey have fought wars in the past and, therefore, Greece might be seen as a potential threat by the Turkish security elite, there is no possibility that Armenia would ever present a military threat to Turkey. Moreover, the EU membership of Greece and the powerful Armenian diasporas in the United States and Europe would most likely nullify the nuisance capability of Turkey's nuclear power against these countries. In addition, Turkey has good neighborly relations with the rest of the countries in its environs, such as Bulgaria, Romania (both now NATO allies), Ukraine, Georgia, and Russia (which has a large nuclear arsenal). All told, it is difficult to conjure a dispute, either internally or in the international community, in which Turkey's use of nuclear weapons would be justified.

There is, therefore, no feasible scenario under which Turkey could expect to effectively use its nuclear power status, if and when achieved. However, there are scenarios in which Turkey's vital interests would be seriously damaged simply because it would have attempted to acquire a nuclear-weapon capability.

Bearing in mind the fact that the "top-secret" meeting of Turkey's foreign minister with top bureaucrats from the military, diplomatic, and intelligence branches of the Turkish state apparatus in his "secure office," where they discussed alternative strategies to deal with the delicate security situation in Syria, leaked to the press on March 27, 2014, keeping a clandestine nuclear-weapon development project away from the watchful eyes of the international community may be next to impossible.

Even though there is talk in Turkey about why Turkey should develop nuclear weapons, among those who approach the issue from the perspective of national pride and prestige as well as security, it is not clear whether they are aware of the possible serious consequences of such action, which would mean, among others, violation of Turkey's international obligations. Secretly going down this path and being discovered could cost Turkey not only international prestige, but also quite possibly NATO guarantees, including the U.S. nuclear umbrella.

NOTES

1   The NPT was signed in 1968 and entered into force in 1970. Mohamed I. Shaker, *The Nuclear Non-Proliferation Treaty: Origin and Implementation 1959–1979*, vols. I, II, and III (New York, London, and Rome: Oceana Publications, Inc., 1980).

2   Yet the NPT could not reach its objective of universality because of the so-called holdouts, such as Israel, India, and Pakistan, which have never joined the treaty, and North Korea, which had joined the treaty in 1985 and walked out in 2003.

3   Stephen M. Meyer, *The Dynamics of Nuclear Proliferation* (Chicago: University of Chicago Press, 1984); Scott D. Sagan and Kenneth N. Waltz, *The Spread of Nuclear Weapons: A Debate* (New York: W. W. Norton, 1995); Scott D. Sagan, "Why Do States Build Nuclear Weapons? Three Models in Search of a Bomb," *International Security* 21, no. 3 (Winter 1996–1997): 54–86; Jacques E. C. Hymans, *The Psychology of Nuclear Proliferation: Identity, Emotions, and Foreign Policy* (Cambridge: Cambridge University Press, 2006); Tanya Ogilvie-White, "Is There a Theory of Nuclear Proliferation? An Analysis of the Contemporary Debate," *Nonproliferation Review* 4, no. 1 (Fall 1996): 43–60; Stanley A. Erickson, "Economic and Technological Trends Affecting Nuclear Nonproliferation," *Nonproliferation Review* 8, no. 2 (Summer 2001): 40–54; T. V. Paul, *Power vs. Prudence: Why States Forgo Nuclear Weapons* (Montreal: McGill-Queen's University Press, 2000).

4   Harald Müller, "The Proliferation of Nuclear Weapons and Missiles: Defense Against Weapons of Mass Destruction," in *Defense Against Weapons of Mass Destruction Terrorism*, ed. Osman Aytac and Mustafa Kibaroğlu (Amsterdam: IOS Press, 2009), 19.

5   Ogilvie-White, "Is There a Theory of Nuclear Proliferation?" 44–48.

6   Müller, "The Proliferation of Nuclear Weapons and Missiles," 20.

7   George C. Reinhardt, *Nuclear Weapons and Limited Warfare: A Sketchbook History* (Santa Monica, Calif.: RAND Corporation, 1964), 8, www.rand.org/pubs/papers/P3011.html.

8   Müller, "The Proliferation of Nuclear Weapons and Missiles," 21.

9   Ali L. Karaosmanoğlu and Mustafa Kibaroğlu, "Defense Reform in Turkey," in *Post-Cold War Defense Reforms: Lessons Learned in Europe and the United States*, ed. Istvan Gyarmati and Theodor Winkler (New York: Brassey's, 2003), 135–64; and Nuray Yentürk, "Measuring Turkish Military Expenditure," *SIPRI Insights on Peace and Security*, no. 2014/1 (March 2014): 1–17.

10 Mustafa Kibaroğlu, "Turkey's Quest for Peaceful Nuclear Power," *Nonproliferation Review* 4, no. 3 (Spring–Summer 1997): 33–44; and Kibaroğlu, "Nuclearization of the Middle East and Turkey's Possible Responses," EDAM Discussion Paper Series, no. 2012/5, Center for Economics and Foreign Policy Studies, November 2012.

11 Two Turkish firms, Sezai Turkes-Fevzi Akkaya (STFA) and TEK, and two Argentine firms, Comision Nacional de Energia Atomica and INVAP, formed the new firm.

12 Argentina would provide Nuclear Steam Supply System (NSSS) technology, basic and detailed engineering for the balance of plant, construction management, and regulatory expertise.

13 "Agreement Signed to Build Carem," *Nuclear Engineering International* (December 1990): 8. Also see Richard Kessler, "Argentina and Turkey to Form Nuclear A/E to Build a Small PWRs," *Nucleonics Week*, October 25, 1990, 12–13. For more on this issue, see "Nuclear Pact With Argentina 'Secretly Signed,'" *Nuclear Developments*, November 15, 1990, 29–30.

14 "Reactor Venture Entered With Turkey," *Nuclear News*, December 1990, 46.

15 Kibaroğlu, "Turkey's Quest for Peaceful Nuclear Power."

16 Author's private conversations and e-mail exchanges with Prof. Dr. Yalcin Sanalan in 1997 for his research paper at the Center for Nonproliferation Studies in Monterey, California, on Turkey's quest for peaceful nuclear energy, and in the subsequent years on the occasion of panels and conferences in Turkey on nuclear energy. Sanalan, who used to be the director of the Turkish Atomic Energy Authority (TAEK) in the 1990s and served as the head of the nuclear engineering department at Hacettepe University in Ankara, gave permission to use this anecdote.

17 Mustafa Kibaroğlu, "The Future of Extended Deterrence: The Case of Turkey," in *Perspectives on Extended Deterrence*, ed. Bruno Tertrais, Coll. Research and Documents, no. 3 (Paris: Fondation pour la Recherche Stratégique, 2010), 87–95.

18 Mustafa Kibaroğlu, "Turkey's Place in the Missile Shield Project," *Journal of Balkan and Near Eastern Studies* 15, no. 2 (Summer 2013): 223–36.

19 International Atomic Energy Agency, "Conclusion of Additional Protocols: Status as of 6 August 2014," www.iaea.org/safeguards/documents/AP_status_list.pdf.

20 "Working Paper Submitted by Belgium, Lithuania, the Netherlands, Norway, Poland, Spain and Turkey for Consideration at the 2010 Review Conference of the States Parties to the Treaty on the Non-Proliferation of Nuclear Weapons," NPT/CONF.2010/WP.69, New York, May 11, 2010, 3, www.reachingcriticalwill.org/legal/npt/revcon2010/papers/WP69.pdf.

21 Other members are Canada, Chile, Germany, Mexico, the Netherlands, Nigeria, the Philippines, Poland, Turkey, and the United Arab Emirates. See "Non-Proliferation and Disarmament Initiative (NPDI)," www.dfat.gov.au/security/npdi.html.

22 Fuat Keyman, "Modernity, Secularism and Islam: The Case of Turkey," *Theory, Culture & Society* 24, no. 2 (2007): 215–34.

23 Hans M. Kristensen, *U.S. Nuclear Weapons in Europe: A Review of Post-Cold War Policy, Force Levels, and War Planning* (Washington, D.C.: Natural Resources Defense Council, 2005), 9.

24 Mustafa Kibaroğlu, "Turkey and Shared Responsibilities," in *Shared Responsibilities for Nuclear Disarmament: A Global Debate*, ed. Scott D. Sagan, Occasional Paper (Cambridge, Mass.: American Academy of Arts and Sciences, 2010), 24–27.

25 Kristensen, *U.S. Nuclear Weapons in Europe*.

26 "Letter to NATO Secretary-General Anders Fogh Rasmussen From the Foreign Ministers of Belgium, Germany, Luxembourg, the Netherlands and Norway," February 26, 2010, www.armscontrol.org/system/files/Letter%20to%20 Secretary%20General%20NATO.pdf.

27 Views expressed by Ambassador Tacan İldem, then director general, International Security Affairs Department, Turkish Ministry of Foreign Affairs, during a one-day workshop convened in Ankara by the Foreign Policy Institute, June 4, 2010. For proceedings of the meeting, see "NATO's New Strategic Concept Conference, June 2010, Ankara," *Diş Politika/Foreign Policy* 36 (Autumn 2010): 9–12.

28 "Active Engagement, Modern Defence: Strategic Concept for the Defence and Security of the Members of the North Atlantic Treaty Organization," adopted by heads of state and government in Lisbon, November 19, 2010, para. 17, www.nato.int/cps/en/natolive/official_texts_68580.htm.

29 North Atlantic Treaty Organization, "Deterrence and Defence Posture Review," May 20, 2012, para. 12, www.nato.int/cps/en/natolive/official_texts_87597.htm?mode=pressrelease.

30 Author's e-mail communication with Professor Tom Sauer, University of Antwerp, Belgium, a scholar known for his work on NATO and U.S. nuclear weapons in Europe, February 24, 2014.

31 Mustafa Kibaroğlu, "The Future of Extended Deterrence."

32 Mustafa Kibaroğlu, "Turkey Says No," *Bulletin of the Atomic Scientists* 59, no. 4 (July/August 2003): 22–25.

33 Mustafa Kibaroğlu, "Missing Bill Clinton," *Bulletin of the Atomic Scientists* 60, no. 2, (March/April 2004): 30–32.

34 Author's recollection from his private conversations with high-ranking Turkish diplomats in the Ministry of Foreign Affairs, Ankara, September 2005.

35 During a TV interview, then foreign minister Ahmet Davutoğlu emphasized that Turkey ranked seventh in the world with its 221 representative missions all over the world and that there is no international organization that Turkey is not represented in. Davutoğlu also argued that as long as Turkey's economic and democratic power continued, Turkey will not retreat from any of the domains to which it belonged. "Davutoğlu'ndan onemli aciklamalar" (Important Remarks by Davutoğlu)," HaberTurk, December 6, 2013, www.haberturk.com/dunya/haber/901141-davutoglundan-onemli-aciklamalar.

36 Davutoğlu is on the record as emphasizing that Turkey will become a "global power" and must therefore adjust its policies and capabilities accordingly. One such statement was made in a speech he gave, while serving as foreign minister, at the "Annual

Ambassadors Conference" held in Ankara on December 23, 2011, www.mfa.gov. tr/disisleri-bakani-sn_-ahmet-davutoglu_nun-iv_-buyukelciler-konferansi-acis-konusmasi_-23-aralik-2011.tr.mfa.

37 Nearly 80 percent of Turks think that Turkey and the United States are no longer allies. The findings of one such poll are at www.transatlantictrends.org.

38 Kibaroğlu, "Turkey's Place in the Missile Shield Project."

39 Tarik Oguzlu and Mustafa Kibaroğlu, "Is the Westernization Process Losing Pace in Turkey: Who's to Blame?" *Turkish Studies* 10, no. 4 (December 2009): 577–93.

40 Antonio M. Ruiz-Jimenez, "European Public Opinion and Turkey's Accession: Making Sense of Arguments For and Against," European Policy Institute Network, Working Paper no. 16, 2007.

41 Lauren M. McLaren, "Explaining Opposition to Turkish Membership of the EU," *European Union Politics* 8, no. 2 (2007): 251–78; Tarik Oguzlu and Mustafa Kibaroğlu, "Incompatibilities in Turkish and European Security Cultures Diminish Turkey's Prospects for EU Membership," *Middle Eastern Studies* 44, no. 6 (November 2008): 945–62.

42 "Turkish PM Faces EU Leaders' Criticism," Al Jazeera, January 22, 2014, www.aljazeera.com/news/europe/2014/01/turkish-pm-faces-eu-leaders-criticism-201412225251993988.html.

43 Kibaroğlu, "Nuclearization of the Middle East and Turkey's Possible Responses."

44 "Scenario" in this case means "a postulated sequence of possible events."

45 "Russia Calls for Full Compliance to Montreux Convention," *Hürriyet Daily News*, August 2008, www.hurriyet.com.tr/english/world/9759360.asp?scr=1.

46 Turgut Özal was prime minister (1983–1989) and president of Turkey (1989–1993).

47 "The Revised Nuclear Suppliers Group Guidelines: A European Union Perspective," Non-Proliferation Papers, EU Non-Proliferation Consortium, no. 15 (May 2012): 7.

48 M. B. Nitikin, A. Andrews, and M. Holt, *Managing the Nuclear Fuel Cycle: Policy Implications of Expanding Global Access to Nuclear Power*, Report for Congress RL34234 (Washington, D.C.: Congressional Research Service, September 12, 2011).

49 U.S. Embassy in Ankara, "Turkey/NSG: Turkey Concerned About 'Subjective Criteria' for ENR Transfers," cable to U.S. State Department, no. 08ANKARA1974, November 14, 2008, http://wikileaks.ch/cable/2008/11/08ANKARA1974.html.

50 "The Revised Nuclear Suppliers Group Guidelines: A European Union Perspective," 9.

51 Fred McGoldrick, *Limiting Transfers of Enrichment and Reprocessing Technology: Issues, Constraints, Options*, Managing the Atom Project, Belfer Center for Science and International Affairs (Cambridge, Mass.: Harvard University Kennedy School of Government, May 2011), 14, http://belfercenter.ksg.harvard.edu/files/MTA-NSG-report-color.pdf.

52 Sinan Ülgen, ed., *The Turkish Model for Transition to Nuclear Power* (Istanbul: Center for Economics and Foreign Policy Studies, 2011), www.edam.org.tr/Media/Files/226/edamreport.pdf.

53 For a detailed discussion of this matter, see Aaron Stein, "Turkey's Missile Programs: A Work in Progress," EDAM Non-Proliferation Policy Briefs (January 2013).

54 Ibid., 6.

55 "Erdogan Cites Iran for Turkish Missile Program," *Hürriyet Daily News*, December
30, 2011; Umit Enginsoy, "Turkey Aims to Increase Ballistic Missile Ranges,"
Hürriyet Daily News, February 1, 2012, www.hurriyetdailynews.com/turkey-aims-
to-increase-ballistic-missile-%20ranges.aspx?pageID=238&nID=12731&NewsCat
ID=345 (cited in Stein, "Turkey's Missile Programs," 6).

56 "Turkey to Work on First ICBM," *Pakistan Defence*, November 23, 2012, http://
defence.pk/threads/turkey-to-work-on-first-icbm.220669.

57 Kibaroğlu, "Turkey's Quest for Peaceful Nuclear Power."

58 Abdul Qadeer Khan headed Pakistan's uranium enrichment program from 1976 to
2001, using centrifuge technology acquired from Urenco, a European uranium-enrich-
ment consortium. In the eyes of most Pakistanis, A. Q. Khan was one of the nation's
greatest scientists, and he rescued Pakistan from the potential domination of a nuclear-
armed India. During his tenure at Khan Research Laboratories, Pakistan's uranium-
enrichment facility at Kahuta, Khan sold gas-centrifuge technology, a type of equipment
that could be used to make enriched uranium for nuclear explosives, to numerous international buyers, including Iran, North Korea, and Libya. For more on this, see Joshua
Pollack and George Perkovich, "The A. Q. Khan Network and Its Fourth Customer,"
January 23, 2102, event summary, Carnegie Endowment for International Peace, http://
carnegieendowment.org/2012/01/23/q-khan-network-and-its-fourth-customer/8vsx.

59 "Turkey's Role in Pakistan's Nuclear Program," *Worldwide Report*, March 20, 1987, 1–4.

60 Mustafa Kibaroğlu, "EURATOM & ABACC: Safeguard Models for the Middle
East?" in *A Zone Free of Weapons of Mass Destruction in the Middle East*, ed. Jan
Prawitz and James F. Leonard (New York and Geneva: United Nations Institute for
Disarmament Research, 1996), 93–123.

61 See "Nuclear Power in Brazil," World Nuclear Association, www.world-nuclear.org/
info/Country-Profiles/Countries-A-F/Brazil.

62 "Ankara 'Adds' Uranium Clause in Nuclear Deal With Tokyo," *Hürriyet Daily News*,
January 8, 2014, www.hurriyetdailynews.com/ankara-adds-uranium-clause-in-
nuclear-deal-with-tokyo.aspx?PageID=238&NID=60729&NewsCatID=348.

63 "Turkey Counters Japanese Media Reports on Uranium Enrichment Option,"
WorldTribune.com, January 14, 2014, www.worldtribune.com/2014/01/14/
turkey-counters-japanese-media-reports-on-uranium-enrichment-option.

64 Mustafa Kibaroğlu and Baris Caglar, "Implications of a Nuclear Iran for Turkey,"
Middle East Policy 15, no. 4 (Winter 2008): 59–80.

CHAPTER 8

DEBATING TURKEY'S NUCLEAR FUTURE

JESSICA VARNUM

W hile this volume's editors graciously answer the question "Why a book about nuclear issues and Turkey?," anyone who watches the news is unlikely to seriously ask such a question. This stands in marked contrast to the immediate post–Cold War era, when only region-alists and North Atlantic Treaty Organization (NATO) experts paid regu-lar attention to Turkey, and most in Western policy circles took Ankara's compliance with Western preferences—on nuclear and non-nuclear issues alike—as an article of faith. As recently as 2009, one reviewer of this author's Turkey case study for an edited volume on proliferation forecast-ing suggested the chapter should be deleted from the volume as Turkey was a "non-puzzle country" and therefore did not warrant the ink.[1]

Six years later, there is no consensus about Turkey's nuclear future, but there is significant debate. Foreign policy luminaries regularly cite Turkey near the top of the list of likely regional proliferation dominoes should Iran acquire nuclear weapons.[2] Researchers have since spilled much ink analyzing the same issues.[3] What follows examines the state of the current debate, in dialogue with other volume authors and the broader literature, and integrates these viewpoints with updates to the author's 2010 proliferation forecasting study of Turkey.[4]

The nuclear proliferation equation is in theory quite simple:

nuclear-weapon intent + capability to acquire nuclear weapons
= nuclear proliferation propensity

Yet solving this equation relies on qualitative estimates rather than quantitative absolutes. How do Turkey's security environment, international relations, and domestic politics increase or decrease its leaders' intent to pursue nuclear weapons? What is the status of Turkey's existing nuclear capabilities? By what means could Turkey acquire nuclear weapons, on what time horizon, and at what direct and indirect costs? Ankara *will* have a nuclear future regardless of whether it pursues nuclear weapons—the question is whether reliance on the NATO nuclear umbrella and adherence to the nonproliferation regime will continue to be pillars of that nuclear future.

TURKEY'S NUCLEAR PRESENT AND FUTURE—THE STATE OF THE DEBATE

There are two main strands to the debate over Turkey's nuclear future. First, there are those who debate Turkey's current nuclear intentions and believe policies such as the AKP (Justice and Development Party) government's adamant pursuit of nuclear energy to be symptomatic of Ankara's intent to develop a latent nuclear-weapon capability. This perspective emanates almost exclusively from U.S. and European policymakers and analysts, whose assessments of Turkey's policymaking are often heavily influenced by mirror-imaging. Faced with badly deteriorating regional security and a latent Iranian nuclear-weapon program, many policymakers' gut reactions are to assume that Turkey will become a proliferation domino. Attempting to put themselves in Turkey's position—but likely still doing so from the standpoint of Western strategic biases—they conclude that any country in Turkey's position would either engage in nuclear hedging or outright pursuit of nuclear weapons. The debate over proliferation dominoes is fairly well-established, however, and is not emphasized in this chapter.

Second, policymakers and analysts debate the trajectory of the evolving conditions that will frame Turkey's future nuclear decisions. Is Ankara's alliance with the United States and NATO strengthening or weakening? Is the Turkish government moving closer to or farther away from domestic, regional, and international policies conducive to nonproliferation? Here there is very little consensus. However, exploring the nuances of the debate over the four variables most relevant to proliferation decisionmaking—Turkey's domestic politics, security environment, multilateral interests and commitments, and current and projected nuclear capabilities—can offer some insights.

DEBATING THE EFFECTS OF DOMESTIC POLITICS ON TURKEY'S NUCLEAR FUTURE

As the late U.S. speaker of the House Tip O'Neill famously observed, "All politics is local."[5] Domestic politics filter security and international politics considerations, shaping Turkish nuclear policy perceptions and decisionmaking.

Broad support exists in Turkey's strategic policymaking community for the country's continued active participation in NATO collective defense. The nuclear-weapon deployments at İncirlik are not widely known about or discussed in Turkey, even by many working in the government. However, as detailed by several authors in this volume, they are perceived by the top civilian and military leadership as tangible evidence of continuing U.S. nuclear deterrence commitments.

Neither the AKP governing party nor any other political party in Turkey overtly advocates developing nuclear weapons. The AKP government's views on nuclear issues are the only ones likely to matter, because civil society remains in its infancy in Turkey. It is unclear—and likely situationally dependent—whether public opinion would support or oppose Turkish development of an independent nuclear deterrent. While there has been some public opposition to nuclear power, proliferation remains a non-issue for the Turkish population.[6] The Center for Economic and Foreign Policy Studies (EDAM) polled a sample of 1,500 Turks concerning their opinions on Turkish acquisition of nuclear weapons, but the wording of the question was highly situational, making the poll's correlation to overall Turkish attitudes about nuclear weapons ambiguous.[7]

Foreign and security policy, including nuclear policymaking, is centralized in the prime ministry. For example, while Turkey's minister of national defense represents Turkey in key bodies such as NATO's Nuclear Planning Group, it is the prime ministry rather than the Ministry of Defense that sets defense priorities. Extensive reforms to civil-military relations under the AKP government have placed the military under civilian control both legally and in practice. Reforms to civil-military relations have, however, created some strategic policymaking vacuums. U.S. policymakers cannot always find relevant counterparts in the AKP government; this is partly because of unfamiliarity with the structure of the new system, but it is also due to the fact that since the elimination of the Turkish General Staff's policymaking role, distinct voids in expertise and authority still exist. Few in the AKP government have responsibility for or expertise in nuclear issues. Key exceptions include personnel charged with implementing the nuclear energy program, principally at the Ministry of Energy and Natural Resources and the Turkish Atomic Energy Authority. However, these individuals do not have responsibility for or expertise in nonproliferation or nuclear-weapon–related issues. Police and border security personnel work to counter illicit nuclear trafficking. Less than a handful of people in the Ministry of Foreign Affairs and the Prime Minister's Office have portfolios with significant nuclear components, and these principally relate to Turkey's engagement with international agreements and organizations such as the International Atomic Energy Agency.

In the long term, it is unclear what the new dynamic in civil-military relations will mean for Turkey's nuclear policies. In the short to medium term, the partial strategic vacuum means that generally, the Turkish government defaults to status quo positions on nuclear-weapon issues, including belief in and reliance on NATO collective defense. While some Western policymakers worry that the Turkish government might consider nuclear weapons if Iran were to acquire a nuclear-weapon capability, or if NATO were to withdraw nonstrategic-nuclear-weapon deployments from İncirlik; there is no evidence to suggest that the Turkish government has had or is having any serious discussions about these or other strategic nuclear issues. Aaron Stein suggests that the bureaucratic separation of Turkey's civil nuclear capabilities from the military and defense ministry complicates the likelihood of nuclear proliferation: "... no Turkish leader has ever sought to leverage its civilian facilities ... for nuclear-weapon–related research. From the outset of Turkey's nuclear era, beginning in 1956 ... the leadership opted to separate the agencies tasked with overseeing NATO nuclear weapons from those entrusted with nuclear energy-related research."[8]

The AKP's unprecedented political success has been possible to an overwhelming degree because of the economic prosperity it has brought Turkey. As such, the AKP government is highly unlikely to pursue nuclear weapons. Decades of economic instability, capped most recently by the 2001 economic crisis, have been followed under the AKP by "steady high growth and modest inflation," as well as decisions by both Moody's and Fitch to declare the country investment-grade.[9] However, Turkey's current account deficit stood at 6 percent of GDP in 2013.[10] As such, the International Monetary Fund "has marked it as one of the most vulnerable economies with external and internal vulnerabilities."[11] Financing this liability will require a continuation of recent foreign investment trends, not a return to pre-AKP levels. Whereas, "[b]etween 1980 and 2000, $10.4 billion of foreign direct investment entered Turkey. In the period 2000–2010, this rose to $100 billion."[12] Given that the AKP's political fortunes are dependent on the country's continued economic prosperity, it cannot risk alienating foreign investors, much less courting international economic sanctions through development of a nuclear-weapon program.

The AKP government faces strong domestic political and economic incentives to keep Turkey's nuclear future firmly linked to NATO deterrence on the one hand and the nonproliferation regime on the other. The principal counterargument to this viewpoint may be the assertion that the current government can no longer be relied upon to look out for its own best interests. Former prime minister Recep Tayyip Erdoğan spoke frequently of an international "interest-rate lobby" seeking to undermine Turkey's leadership and political stability.[13] He also vocally sought observer membership in the Shanghai Cooperation Organization and played a particularly dangerous game of chicken with NATO over the possible acquisition of a Chinese national missile defense system. Such behavior was perceived by many in the United States and Europe as erratic and self-defeating. It contributed to the end of the special Obama-Erdoğan friendship, calling into question for many the overall predictability of Turkish policymaking.[14]

Others argue, however, that Erdoğan's behavior represented a calculated play to populist politics, or as Michael Koplow framed it, a decision to sacrifice "... Turkish influence internationally in order to solidify his [Erdoğan's] position at home."[15] If Koplow is correct that the prime minister was co-opting Turkey's foreign policy in service of domestic populist politics, this is also a worrying trend. The dramatic increase in Turkish leaders' "foreign conspiracy" rhetoric highlights resurgent nationalist and isolationist tendencies, moving Turkish politics away from the internationalism of the early AKP government. Examining debates over Turkey's security environment and multilateral commitments clarifies the context for this trend.

DEBATING THE EFFECT OF TURKEY'S SECURITY ENVIRONMENT ON ITS NUCLEAR FUTURE

Turkey's perception of the threats to its security and its capacity—alone or with allies—to deter or prevail over likely adversaries is a critical variable in its nuclear policymaking. Do Turkish elites view nuclear weapons as useful vis-à-vis the country's principal security threats, and if so, does U.S./NATO extended deterrence adequately meet their needs?

In recent years Turkey has primarily focused on non-state actor threats that cannot be solved through nuclear deterrence. In a new Middle East defined by brutal civil wars and fractured states, Ankara faces, as former president Abdullah Gül put it, threats such as "Syria becoming Afghanistan on the Mediterranean."[16] Al-Qaeda affiliates perpetrated car bomb attacks in Reyhanlı in May 2013 and remain an ongoing threat.[17] The Islamic State controls or is seeking to control much of the territory on Turkey's Iraqi and Syrian borders, and held 46 Turkish diplomatic personnel hostage for more than three months. Moreover, with the unravelling of the Turkish government's peace process with the separatist Kurdistan Workers' Party (PKK) it appears the principal threat to Turkey's security of the past two decades may remain ongoing.[18]

As recently as 2010 Turkey's government perceived its regional security environment to be free of any serious (state-based) threats, largely because current strategic thinking emphasizes neighbors' intentions rather than their capabilities.[19] Despite its overall focus on non-state threats, however, Ankara has clearly remained concerned since the 1991 Gulf War about both immediate and longer-term asymmetric threats from neighbors possessing weapons of mass destruction. As neighboring regions, from the Middle East to the Caucasus, succumb to civil war, revolution, or the "unfreezing" of territorial disputes, Michael Koplow notes that Ankara "… has gone from a zero problems with neighbors policy to one in which it is hard to find any former regional ally left with whom Turkey is not feuding to one degree or another."[20] Turkey no longer has an ambassador in either Egypt or Israel and frequently finds itself at odds with Iraq's central government over its increasingly robust partnership with Erbil's Kurdistan Regional Government. Formerly warm relations with Iran and Russia are strained over Turkey's decision to side with NATO on missile defense and its anti-regime stance in the Syrian civil war. Low-level hostilities between Turkish and Syrian regime forces on their shared border, combined with the Turkish government's desire for regime change in Syria and support for rebel factions, frequently threaten to drag Ankara into more direct conflict with Assad's forces. Meanwhile, Ankara is also unwillingly in the middle of Western and Kurdish efforts to destroy the Islamic State, knowing that arming the Kurds to fight against it could fuel future PKK attacks.

Although the possibility of a nuclear Iran remains an ongoing concern, it appears to be a very low-priority threat for both Turkey's general public and its foreign policy experts. A fall 2013 poll by EDAM found "that the biggest threat perceived by the Turkish public remains the foundation of an independent Kurdish state in [the] southern part of Turkey. For the expert community however the biggest threat is the dominance of Islamist extremists in Syria."[21] Both viewed "military intervention by the U.S. or Israel in order to prevent Iran from getting nuclear weapons" as a greater threat to Turkey than "Iran gaining nuclear weapons."[22] As of late 2014, relative to other regional issues, the Iranian nuclear program is simply not on Turkey's agenda.

While direct military hostilities between Russia and Turkey are highly unlikely (both countries still perceive their relationship as cooperative), a regionally resurgent nuclear-armed Russia may pose one of the most serious security dilemmas for Turkish policymakers. Turkey and Russia continue to have competing regional interests, and they support opposing factions in nearly every neighboring conflict of interest (for example, Syria, Egypt, and the "frozen" conflicts over Nagorno-Karabakh, South Ossetia, Abkhazia, and Transnistria). Russia's forcible incorporation of Ukraine's Crimean Peninsula in March 2014 (preceded in 2008 by a war with Georgia over South Ossetia and Abkhazia) signaled a worrying renewed Russian activism in Turkey's immediate neighborhood, often involving countries with large ethnically Turkic minorities.[23]

What are Turkey's capabilities against the security challenges it faces? Turkey's conventional military capabilities are among the most formidable in Europe and the Middle East, thanks in part to decades of military cooperation with the United States. In the context of combating both PKK and Islamist terrorism, and fending off spillover from the Syrian civil war, Turkey's defense and procurement policies have emphasized modern conventional capabilities, from helicopters to drones, and the country has the second-largest army among NATO member states.

Ankara's first line of deterrence, and if necessary defense, against state threats is the U.S./NATO security guarantee. A member of NATO since 1952, Turkey was a geostrategically high-value Cold War ally that still hosts 60–70 U.S. nonstrategic nuclear weapons at İncirlik Air Base as part of NATO's extended deterrence nuclear umbrella.[24] As demonstrated by the 2010 Lisbon summit's decision to situate the Allied Land Command in İzmir when NATO consolidated its headquarters from eleven locations to six, and the decision to base NATO missile defense's X-band radar in Kürecik, Turkey continues to be centrally important to the alliance. Yet Ankara remains concerned about the credibility of NATO security guarantees in various contexts. While six NATO PAC-3 missiles, deployed in Patriot batteries, and 1,200 associated NATO personnel have been deployed near Turkey's Syrian border since January 2013, Turkish policymakers still view the alliance's collective defense track record as spotty. For example, Doruk Ergun writes in chapter 3, "… the alliance was slow to deliver Patriot anti-missile batteries during the Gulf War, when the risk to Turkey was urgent."

The limits of Turkey's faith in alliance commitments become even clearer when placed in the context of the debate over whether the United States should continue to station tactical nuclear weapons in Europe, and if NATO missile defense, once completed, might provide a viable strategic alternative to the weapons deployments. In chapter 4, Can Kasapoğlu argues that "… the essence of ballistic missile defense systems and tactical nuclear weapons is different from a military standpoint. Thus, they would be best used in harmony rather than compensation…. the nature of ballistic missile defense has very little capability in retaliatory action, coercive and gunboat diplomacy endeavors, preemptive military missions, and projecting active deterrence." The premise of Kasapoğlu's entire debate over tactical nuclear weapons versus ballistic missile defense appears to be shared by many Turkish policymakers, but it gets to the heart of why U.S. counterparts may fail to appreciate the strategic importance to Turkey of the tactical-nuclear-weapon deployments. Certainly, it is true that missile defense is not a strategic substitute for nuclear weapons. But from the U.S. perspective, NATO extended deterrence is underpinned by U.S. strategic nuclear weapons. Removing tactical-nuclear-weapon deployments from NATO allies' territories therefore would not be a case of "substituting" missile defense for nuclear weapons. In the unlikely event that the United States needed to use nuclear weapons to make good on its extended deterrence commitments, it is all but inconceivable that it would use any of the tactical nuclear weapons currently stationed in Europe. Speaking of the B61, General James Cartwright, former vice chairman of the Joint Chiefs of Staff, asserted: "Their military utility is practically nil…. They do not have assigned missions as a part of any war plan."[25]

But for Turkey, militarily useless nuclear weapons stationed on its territory may continue to mean more than the robustly capable U.S. strategic nuclear weapons deployed outside of its territory. Partly, this is because the political arrangements underpinning the weapons deployments institutionalize Turkish decisionmaking participation in ways likely to become obsolete if all nuclear deterrence functions default to U.S. (and possibly French and British) strategic systems. NATO nuclear training missions currently involve allied forces, even if only in support roles. NATO countries participate in the Nuclear Planning Group, which as Doruk Ergun observes, "…allows Ankara to influence NATO's nuclear posture while building alliance cohesion and promoting burden sharing by including non-nuclear NATO members in the process."

Tactical nuclear weapons are also symbolically critical, however. For decades, Turkish policymakers have questioned U.S. willingness to trade New York for Istanbul in any nuclear contingency. Such doubts have only intensified since the end of the Cold War, as it is unclear whether, in practice, NATO's security guarantees actually extend to Middle East contingencies. As Can Kasapoğlu observes, "… using these nuclear bombs in a conflict would strictly depend on a unanimous decision within NATO.…" The credibility of NATO is fundamentally about the credibility of the United States. To what degree does that credibility rest on the continued deployment of U.S. nuclear weapons in Turkey? As of late 2014, it seemingly remains the case that the weapons are a "symbolically—though not militarily—critical component of the bilateral relationship," in that "leaders have traditionally considered them to be tangible evidence of the U.S. commitment to Turkey's defense."[26]

Tangible commitments are important to Turkey because, as Joshua Walker puts it, "Turkish children are taught, even before they know how to speak, the expression 'Turks have no friends except for other Turks.'"[27] This, from the Turkish perspective, has been repeatedly reinforced in relations with the West. For example, Turkey's insistence that military procurement contracts include extensive co-production and technology transfer provisions is about a drive for self-sufficiency that traces back to the four-year arms embargo the United States placed on its military ally in 1975 after Turkish military action in Cyprus. While incidents such as the embargo are ancient history for Americans, they deeply resonate in Turkish policymaking to this day. Murat Karagöz observes that the embargo "created, especially in the eyes of the Turkish public, a deep lack of confidence towards the United States."[28]

It appears, however, that a resurgent nuclear-armed Russia combined with broader regional instability is moving Turkish policymakers closer to traditional Western allies. In response to the extreme displeasure of its NATO partners over its 2013 selection of a Chinese national missile defense system, Turkey appears likely instead to eventually pursue a deal with French-Italian Eurosam, or at the very least to avoid finalizing a deal with China.[29] And Turkey is moving in concert with its Western allies on such regional challenges as the Crimean Peninsula. According to prominent Turkish journalist Semih İdiz, "… the 'revenge of geography,' to use

Robert D. Kaplan's term, is forcing Ankara to return to Turkey's tradition-ally cautious diplomacy, based on a preference for multilateralism, while it maintains and deepens security arrangements with the West."[30]

DEBATING THE EFFECTS OF MULTILATERAL INTERESTS AND COMMITMENTS ON TURKEY'S NUCLEAR FUTURE

Beyond collective security organizations such as NATO, what are Turkey's primary multilateral interests and commitments? How are they likely to shape its nuclear future?

Turkey is a party to the Treaty on the Non-Proliferation of Nuclear Weapons (NPT), and all of the major nonproliferation treaties and organi-zations, as well as a participant in numerous voluntary arrangements. As discussed by Mustafa Kibaroğlu in chapter 7, many nonproliferation experts worry that the NPT is close to a tipping point; if Iran or any other non-nuclear-weapon states parties to the treaty acquire nuclear weapons (as North Korea did), the regime could lose all remaining credibility, prompt-ing countries like Turkey to feel they should no longer be bound by the regime. Turkey's belief in the universal and reciprocal nature of the regime makes such scenarios worthy of concern. For example, former president Gül recently asserted: "It is a contradiction to consider attempts by some countries to seek chemical and biological weapons because they are cheap and easy to attain to be illegitimate while possessing highly lethal and sophisticated weapons, including nuclear arms, is seen to be legitimate."[31]

Yet Turkey's position vis-à-vis NPT implementation is far from straightforward. As Mark Hibbs observes in chapter 6, "The reality of Turkey's security arrangements [reliance on NATO extended deterrence] has deterred Ankara from going beyond disarmament rhetoric." Unlike many other non-nuclear-weapon states, Ankara's faith in the NPT regime is not dependent on the nuclear-weapon states' intensifying their nuclear disarmament efforts under Article VI of the treaty. However, Turkey sides with many of the non-nuclear-weapon states in its interpretation of the NPT's peaceful uses clause (Article IV). As a member of the Nuclear Suppliers Group, Ankara has strongly advocated that all non-nuclear-weapon states in compliance with the NPT be guaranteed the right to possess peaceful nuclear technologies, including enrichment and reprocessing capabilities. Further curtailment by nuclear-weapon states of what Turkey and like-minded non-nuclear-weapon states view as their "inalienable rights" in regard to peaceful nuclear technology would pose a serious threat to the continued credibility of the NPT.

Diplomatic developments vis-à-vis Iran's enrichment program bear close monitoring in this regard. Victor Gilinsky and Henry Sokolski argue that if finalized, the Joint Plan of Action with Iran could set a dangerous precedent that the international community views uranium enrichment by non-nuclear-weapon states as acceptable, and thus "... is not just about Iran. It is about the rules for nuclear power programs throughout the world."[32] Predicting the effects of an Iran settlement that permits some level of enrichment is not that straightforward, however. Most non-nuclear-weapon states are interested in upholding the principle but are unlikely to pursue enrichment technologies because they are prohibitively expensive and complex. Additionally, the nuclear suppliers are unlikely to provide new states with enrichment capabilities. None have done so in recent decades; rather, countries such as Iran and North Korea obtained centrifuge technology illicitly from Pakistan's A. Q. Khan network. Supplier states have exercised restraint in part because of tighter non-proliferation rules, but more fundamentally, they face strong incentives to keep new enrichment services competition out of the market.

Turkey's long-term commitment to the nuclear nonproliferation regime is also inextricably linked to its continued acceptance of the legitimacy of international institutions and multilateralism. Multilateralism is a relatively new phenomenon in Turkish political culture, prominently developing only since Ankara became a member of NATO in 1952. And Turkey's newly assertive foreign policy is even more recent, a product of the AKP government's desire to take a more active role in shaping Turkey's regional and global neighborhood. Aaron Stein observes that Turkey's government and citizenry are becoming disenchanted with international engagement, noting that the government's "failures to alter the course of the [Syrian] conflict began to be seen as Turkish impotence.... this reignited the calls within the country to retreat from its more ambitious foreign policy to the Kemalist favored non-interventionist 'peace at home, peace abroad' foreign policy."[33] While the blowback Turkey has experienced in regard to its policies in Syria, Egypt, and elsewhere is at least partly attributable to

Turkish miscalculations, the government has not hesitated to place the blame squarely on other countries and multilateral groupings such as NATO and the UN Security Council. Former prime minister Erdoğan vocally called for UN Security Council reform, arguing in the context of Russian and Chinese opposition to intervention in Syria that "[t]he U.N. has made us question its own existence during the whole conflict.... Imprisoning the U.N. to what the five permanent members are going to say is anti-democratic.... Youth have started a campaign named the world is bigger than five. I support it."[34]

It is in this context—maintaining Turkey's belief in and commitment to multilateralism—that levers less specifically about the nonproliferation regime, namely European Union membership, become critically important. The AKP government came to power in 2002 on a mandate that included landslide support for Turkish EU membership. Using the EU process as leverage, the government successfully liberalized key aspects of Turkey's economy and political structure, striking a death blow to decades of military tutelage. The European Union officially commenced the accession process with Turkey in 2005 (eighteen years after Ankara first sought membership); nine years later, negotiations have repeatedly stalled and the parties are at an impasse. A 2013 German Marshall Fund poll indicates that "popular support in Turkey for EU membership has fallen to 44 percent from 73 percent in 2004."[35] Formerly supportive Turkish policymakers play populist politics by spouting anti-EU rhetoric. Turkey's former EU affairs minister Egemen Bağış asserted: "Turkey doesn't need the EU, the EU needs Turkey. If we have to, we could tell them, 'Get lost, kid!'"[36]

Meanwhile, Turkey has (unsuccessfully) sought observer membership in the Shanghai Cooperation Organization, with officials regularly citing it as an alternative to EU membership. In January 2013, Erdoğan argued, "When things go so poorly, you inevitably, as the prime minister of 75 million people, seek other paths. That's why I recently said to Mr. [Vladimir] Putin: 'Take us into the Shanghai Five; do it, and we will say farewell to the EU, leave it altogether. Why all this stalling?'"[37] While Turkey's rhetoric invoking the Shanghai Cooperation Organization has angered Western partners and worried analysts, it lacks substance. The Shanghai Cooperation Organization remains highly limited in its ability to deal with either the security or economic concerns of its members, given that it is led by two rival powers—Russia and China—whose interests rarely align sufficiently to allow for meaningful action on anything. Russia and China also both face strong incentives to exclude Turkey—a veritable Western Trojan horse—from attaining more than its current "dialogue partner" status.

Former Turkish president Gül took a far more conciliatory tone, calling the European Union the "'centre of gravity' of Turkish foreign policy."[38] And in November 2013, after three years of deadlock, the European Union opened another chapter in the accession negotiations. Because this was followed later in 2013 and early 2014 by yet more scandals—including the Turkish government's decision to sharply curtail civil liberties by blocking access to Twitter and YouTube—prospects for serious movement on EU membership seem dimmer than ever. As such, the question is whether an end or perceived end to Turkish EU membership prospects is likely to influence Turkey's future nuclear policies. Both Turkish and Western experts have argued that EU membership would provide additional long-term incentives for Turkey to forgo nuclear weapons by tangibly anchoring it in the West.[39] However, most—including this author—identify NATO membership and the underlying bilateral relationship as the far more important factor.

Proliferation theory suggests that a state's international integration correlates to some degree with its nuclear proliferation propensity.[40] Ankara's unprecedented international economic integration is perhaps the lever of most interest in preventing a retreat from multilateralism. As long as the country remains interdependent with the European economic zone, the status of Ankara's EU membership bid is unlikely to be the determinative variable in Turkey's relationship with Europe. Similarly, ongoing global economic integration is likely to help reinforce Ankara's ideological commitment to the nonproliferation regime.

DEBATING THE EFFECTS OF CURRENT AND PROJECTED CAPABILITIES ON TURKEY'S NUCLEAR FUTURE

The last debate of major interest concerns Turkey's current and projected capabilities to develop nuclear weapons indigenously, and the degree to which these are likely to shape nuclear policymaking. The dual-use technology most relevant to nuclear weapons includes requisite fissile materials and conventional components, indigenous technical expertise, and related delivery systems. As detailed in this section, Turkey's basic nuclear research infrastructure and expertise would provide it with minimal advantages, compared with a country that lacks nuclear infrastructure, if it decided to produce fissile materials for weapons. Similarly, its indigenous abilities to design an appropriate missile delivery system are highly limited.

Both the conventional wisdom and many experts contend that in general, a civil nuclear program is likely to increase a country's proliferation propensity—though in some cases, perhaps by only a negligible amount. Political scientist Matthew Fuhrmann argues that "countries that receive peaceful nuclear assistance are more likely to initiate weapons programs and successfully develop the bomb, especially when they are also faced with security threats."[41] More specifically, many experts argue that a number of Middle Eastern and Asian states—including Turkey—are pursuing nuclear power programs primarily as a nuclear-weapon hedging strategy, developing capabilities and expertise that would enable them to fast-track a future nuclear-weapon decision.[42] Both assumptions are poorly technically grounded, as the most challenging stages of a nuclear-weapon program—uranium enrichment and/or plutonium reprocessing—are not included in most countries' civil nuclear programs generally, or Turkey's plans specifically. Turkey currently lacks any experience with enrichment or reprocessing technologies.[43] A country with Turkey's nuclear inexperience (it plans to rely on both of its nuclear suppliers to operate its first two nuclear power plants), would require significant foreign assistance to embark on either fuel-cycle technology. Pursuing these capabilities would cost far more than the already unaffordable planned nuclear power plants, invite international opprobrium, and derail Ankara from realizing the very real energy independence objectives of its nuclear power program.

More fundamentally, despite a flurry of inaccurate media reports to the contrary in early 2014, Turkey does not appear to have any near-term plans to pursue enrichment or reprocessing technologies. The *Asahi Shimbun* alleged that Ankara's pending nuclear cooperation agreement with Japan contained "...a provision that would enable Turkey to eventually enrich uranium and extract plutonium by reprocessing spent nuclear fuel."[44] The claim was soon widely republished as fact by media outlets in Turkey and worldwide, despite the reality that the nuclear cooperation agreement contains standard prior consent provisions that would require a revised agreement to be concluded with Japan for enrichment or reprocessing of Japanese-origin materials to take place.[45] Additionally, Turkish Energy Minister Taner Yıldız went on the record in January 2014 to clarify: "We don't have any project regarding nuclear fuel and enrichment," further indicating that Turkey does not plan to enrich uranium in the future.[46] As noted both earlier in this chapter and by Mark Hibbs in chapter 6, the Turkish government has taken a strong principled stand concerning the rights of non-nuclear-weapon states in compliance with their NPT commitments to enrichment and reprocessing technologies. The Turkish government would also like to keep open all long-term options, in case it ever did change its mind about enrichment. As Hibbs notes: "A senior Turkish official said that Turkish sensitivity about U.S. advocacy of efforts to limit enrichment and reprocessing may have been heightened by the historical memory of a U.S. arms embargo imposed after Turkey invaded Cyprus in 1974." For the present, however, Turkey is capable of pursuing neither near-term enrichment or reprocessing capabilities, nor does it appear to be interested in doing either.

Arguably, successful implementation of the AKP's nuclear energy program would further constrain Turkey from pursuing a nuclear-weapon program. Because the AKP government is pursuing the program in a fashion that is overwhelmingly dependent on the ongoing goodwill of foreign suppliers, it cannot afford to risk their alienation. Certainly, some countries have used imported "peaceful" nuclear technology in service of weapon programs, but they have done so at a high cost, at least temporarily impairing their nuclear energy programs. Were Turkey to withdraw future foreign-supplied power reactors from safeguards in service of a nuclear-weapon program, no legitimate nuclear supplier would continue to provide it with expertise, technology, or enriched uranium fuel. And this leads to a more fundamental reality; given its rapidly increasing energy demand and astounding 72 percent energy import dependence, Turkey genuinely needs both nuclear power and ready access to oil and gas from foreign suppliers.[47] As such, Turkey's nuclear energy program will arguably lessen, at least in the short and medium term, the likelihood that the country will consider a nuclear-weapon program. The greatest proliferation risks associated with the program concern the long-term possibility that the country's domestic political landscape and alliances radically change, but if this were to happen the international community would likely face bigger challenges than the presence of light-water reactors on Turkish soil.

Sophisticated weapon design capabilities (including nuclear warhead miniaturization) and development or acquisition of a suitable ballistic or cruise missile delivery vehicle would also be required for a modern nuclear-weapon program.[48] Indigenous missile work in Turkey to date has been extremely basic. In chapter 5, Aaron Stein comments on Turkey's ambitions to design a longer-range cruise missile, as well as a rocket that could serve as a satellite launch vehicle and/or the basis for the country's first ballistic missile. But Ankara's only design experience involves the standoff precision-guided cruise missile, and cruise missiles are fundamentally different technologies than ballistic missiles or satellite launch vehicles. Additionally, the Turkish government's stated missile development ambitions are limited by its perpetual procurement overstretch, both in terms of financing and domestic follow-through capabilities. While the government has sought ballistic missile defense systems, drones, new attack helicopters, and a lengthy list of other advanced military hardware items, none of these major projects has yet produced results.

Expertise is another critical dimension of a country's capabilities. Turkey's lack of nuclear expertise contributes to an overall lack of discussion of nuclear issues that reinforces a continued lack of expertise. Few Turkish elites appreciate the complexities of nuclear-weapon development. As Henri Barkey observed in 2009, "Discussion in Turkey has remained conjectural and, with few real specialists on the subject, has had a somewhat unreal quality to it."[49] For example, in 2007 Seyfi Taşhan, director of Ankara's Foreign Policy Institute, addressed whether or not Turkey would "have the capacity to make a bomb now." Taşhan said, "We have two research reactors, rich uranium mines, hundreds of scientists and the most developed industrial and technological infrastructure in the Middle East. In fact, all that we are lacking is the fissile material."[50] This assessment dramatically underestimates the effort that would be required to produce the requisite fissile materials, design a functioning weapon, and pair that weapon with an appropriate delivery system. As Scott Sagan compellingly argues concerning some historical overestimates of countries' proliferation capabilities: they "have too often used misleading measures of the key variables involved in nuclear technology, focusing on broad measures of industrial capability and nuclear research reactor experience and not on the specific fuel-cycle technologies and facilities needed to make the fissile materials required for a nuclear weapon."[51]

Certain AKP government policies are likely to exacerbate even further the gap between expertise and policymaking. For example, decree laws 643 and 649 undermine Turkey's independent regulatory authorities by eliminating most of their financial and decisionmaking autonomy.[52] İzak Atiyas extensively discusses in chapter 2 why it is critical for Turkey to establish an independent nuclear regulatory authority. Atiyas mentions some challenges associated with human capital; this author's 2013 study found inadequate indigenous expertise to be one of the principal challenges facing Turkey's nuclear energy program.[53] The institutional incapacity of independent regulatory authorities to select personnel based on merit—because they are not truly independent—is likely to prevent such expertise from being successfully cultivated by the Turkish government.

Some might argue that, in light of Turkey's poor nuclear expertise, a high-level decision to pursue nuclear weapons could be made absent meaningful assessment of the country's true capabilities; this is arguably exactly what happened when the Turkish government selected first-of-their-kind nuclear power plants completely inappropriate to Turkey's regulatory limitations.[54] This would, however, be highly unlikely in the near term given top leadership's consistent rhetorical support for nuclear non-proliferation, as outlined by Mustafa Kibaroğlu in chapter 7. Lack of expertise does, however, frustrate dialogue between the United States and Turkey on issues ranging from the facts on the ground in Iran to non-strategic nuclear weapons.

Overall, Turkey's limited nuclear capabilities and expertise are likely to reinforce status quo nuclear policies. Already, the government has made capabilities procurement decisions that signal a continued intent to rely on NATO extended deterrence. Unlike Belgium, the Netherlands, and Germany, which have not committed to replacing their dual-capable aircraft when existing capabilities are phased out around 2025, Turkey plans to procure F-35s compatible with the NATO nuclear mission.[55] Meanwhile, the U.S. government is in the process of modernizing the B61-12 gravity bombs stationed in Europe with the addition of a guidance system tail kit, and plans to integrate the new systems with NATO aircraft.[56] Because the new system will greatly improve the accuracy of the weapons, modernization may ironically make the weapons useful—and therefore usable—from a military standpoint. It remains too soon to determine what impact, if any, this might have on NATO and U.S. nuclear war planning, strategic stability vis-à-vis countries such as Russia, and perceptions by allies such as Turkey concerning the importance and credibility of the nonstrategic-nuclear-weapon deployments.

CONCLUSIONS ON TURKEY'S NUCLEAR FUTURE

There is significant potential value to exercises like the one undertaken in chapter 7, which questions the "long-term reliability of the factors that kept Turkey from acquiring nuclear weapons." Just as Mustafa Kibaroğlu looks at possible "worst-case scenarios" in which the Turkish government might consider pursuing nuclear weapons, this author's 2010 publication[57] examined "trends and trigger events" that could alter Ankara's nuclear proliferation decisionmaking.

Such exercises have policy utility to the degree that they build understanding of the factors most relevant to specific countries' proliferation decisionmaking, enabling policies of prevention or treatment. For example, U.S. policymakers frequently question which U.S. actions (or inactions) might contribute to n-country proliferation scenarios. Would withdrawing NATO tactical-nuclear-weapon deployments cause countries like Turkey to lose faith in extended deterrence? Would a nuclear-armed Iran precipitate a chain reaction of proliferation decisions among neighboring countries? Many such scenarios could influence the direction of Turkish nuclear policymaking. However, there appears to be a strong presumption in both this volume and the broader literature that only rarely could a single "trend or trigger event" cause a country otherwise predisposed to nonproliferation to actually launch a nuclear-weapon program. In Turkey's case, a range of factors, including domestic politics, the security environment, multilateral interests and commitments, and current and projected nuclear capabilities, favor continued proliferation restraint.

Capabilities limitations—and more precisely, the political and economic stability trade-offs required to overcome these limitations—are likely to significantly constrain the Turkish government from reaching any near-term "tipping points" as a result of individual events that negatively affect the other political-strategic variables discussed. As noted in the capabilities section, Turkey is far from possessing and unlikely to pursue the enrichment or reprocessing capabilities necessary for a serious nuclear-weapon program. Furthermore, Turkey is unlikely to pursue the covert procurement options discussed in chapter 7.[58] As such, any future proliferation decision would likely require Turkey to pursue enrichment or reprocessing capabilities under the guise of a peaceful program, a risky and time-consuming endeavor.

Unless the nonproliferation landscape fundamentally changes after a hypothetical final deal with Iran that permits some level of enrichment, many of the nuclear-weapon states, Turkey's Western allies, and other countries in Turkey's region would almost certainly find a Turkish enrichment or reprocessing program extremely worrisome. The United States in particular would likely attempt to persuade or coerce Turkey into giving up any enrichment or reprocessing ambitions, as it continues to do vis-à-vis both existing and prospective nuclear cooperation partners, such as South Korea. However, any Iran deal permitting enrichment may forcibly change the U.S. approach to new enrichment programs in non-nuclear-weapon states by setting a troubling precedent. As such, the effects of an Iran deal on the overall nonproliferation regime are one of the great unknowns for those forecasting Turkey's nuclear future.

Current and probable near-term Turkish governments are also unlikely to tolerate the reputational costs associated with the pursuit of nuclear weapons, as Turkey does not self-identify as a pariah state. If anything, the current government has coveted regional and international influence more than any recent predecessors. Mustafa Kibaroğlu argues in chapter 7 that Turkey's persistent participation and leadership in nonproliferation initiatives in recent decades has also made the country less likely to decide to develop its own nuclear weapons because it has helped to create "...a cadre of civil and military bureaucrats, scholars, scientists, experts, and intellectuals who have...a high degree of consciousness about the possible consequences of developing nuclear weapons clandestinely."

Domestic politics also favor status quo nuclear policies. In chapter 1, Gürkan Kumbaroğlu attributes the fact that "Turkey has managed to make much more headway [on nuclear energy] since 2010 than in the five decades preceding it" to the country's dramatically improved political and economic stability. Altering its present course to pursue nuclear weapons would not only involve trade-offs for Turkey with the nuclear energy program; it would also shatter the (relative) political and economic stability the country has enjoyed in recent years. Turkey is not a country, nor is the AKP government a regime that is willing, like Pakistan or North Korea, to "eat grass" in order to obtain nuclear weapons.[59] As Soner Cagaptay observes, "... Turkey has become a majority middle-class society, the first such society in Muslim history."[60] Its citizens, and AKP voters particularly, are interested in preserving their newfound economic well-being.

The author's 2010 forecasting study observed: "Any event that dramatically worsens Turkey's regional security environment could increase its proliferation incentives."[61] Yet, as outlined earlier, Turkey still does not face any threats that would be ameliorated by the acquisition of an independent nuclear deterrent. Rather, pursuing nuclear weapons would render Ankara less secure, by damaging its ties with allies and heightening tensions with other states in the region. While it is useful to consider the circumstances under which this might change—and whether NATO's nonstrategic nuclear weapons are likely to play an important role in preventing such scenarios—Turkey's current balance of threats and capabilities suggests no near-term benefit to pursuing nuclear weapons.

Generations of medical students have been taught "when you hear hoofbeats, think horses, not zebras." That is, the simplest explanation for a patient's symptoms is often the correct explanation. Similarly, if exotic conditions are required for a country to pursue nuclear weapons, the presumption should be that it is unlikely to do so. Turkey's strategic orientation remains of ongoing policy interest because it is a rising power whose choices will shape its regional neighborhood and beyond. As such, future studies will and should be written concerning whether evolving circumstances suggest Ankara might pursue nuclear weapons. As of late 2014, however, analysis of both the factors underlying Turkish nuclear decisionmaking and the overall state of the expert debate suggest that Turkey's nuclear future— at least in the short to medium term—will remain a peaceful one.

NOTES

1   Specifically, the reviewer commented, "I suggest deleting this chapter. As far as I know Turkey has never contemplated going nuclear. It is a member of NATO protected by NATO nukes under article 5. It badly wants to join EU. So it is not a puzzle why it has not gone nuclear.... It is also excessively lengthy for a non-puzzle country."

2   For example, see Geneva Sands, "Clinton: 'I Can Absolutely Bet' a Nuclear Iran Would Spark a Regional Arms Race," *Hill*, June 21, 2012, http://thehill.com/video/administration/234131-clinton-i-can-absolutely-bet-a-nuclear-iran-would-spark-a-regional-arms-race#ixzz2hfOmMQnw; Matthew Grant Anson, "George Mitchell: Nuclear Iran Would Ignite Worldwide Arms Race," *American News Report*, March 2, 2012, http://americannewsreport.com/nationalpainreport/former-sen-mitchell-nuclear-iran-would-ignite-worldwide-arms-race-8813376.html; Jeffrey Goldberg, "Obama to Iran and Israel: 'As President of the United States, I Don't Bluff,'" *Atlantic*, March 2, 2012, www.theatlantic.com/international/archive/2012/03/obama-to-iran-and-israel-as-president-of-the-united-states-i-dont-bluff/253875.

3   For example, see Sinan Ülgen, "Turkey and the Bomb," Carnegie Paper, Carnegie Endowment for International Peace, February 2012. See also Christopher Hobbs and Matthew Moran, *Exploring Regional Responses to a Nuclear Iran: Nuclear Dominoes?* (New York: Palgrave Macmillan, 2014).

4   Jessica C. Varnum, "Turkey in Transition: Toward or Away From Nuclear Weapons?" in *Forecasting Nuclear Proliferation in the 21st Century: Volume 2, A Comparative Perspective*, ed. William C. Potter and Gaukhar Mukhatzhanova (Stanford, Calif.: Stanford University Press, 2010).

5   Tip O'Neill and Gary Hymel, *All Politics Is Local: And Other Rules of the Game* (B. Adams, 1995).

6   Pınar Tremblay asserted, "At this point [nuclear proliferation] is a non-issue for the public." Telephone interview with Pınar K. Tremblay, University of California, Los Angeles, June 23, 2008. Quoted in Varnum, "Turkey in Transition: Toward or Away From Nuclear Weapons?"

7   The polling question was framed in terms of an assumed threat from a nuclear-armed Iran, while some might argue that a nuclear Iran would not inherently represent a threat to Turkey. Respondents were also given only two options—reliance on NATO or Turkish nuclear weapons. As such, the data may reflect NATO skepticism on the part of the general public more than they reflect public support for nuclear-weapon development. The poll asked, "In reaction to a possible threat from a nuclear-armed Iran, should Turkey develop its own nuclear weapons or rely on NATO's protection?" See "Conditional Support for Nuclear Armament," in *Public Opinion Surveys of Turkish Foreign Policy* 2012/1, Center for Economics and Foreign Policy (EDAM), 2012; "54 Pct of Turks Support Nukes If Iran Has Them," *Hürriyet Daily News*, March 29, 2012, www.hurriyetdailynews.com/54-pct-of-turks-support-nukes-if-iran-has-them.aspx?pageID=238&nid=17151.

8   Aaron Stein, "Turkey's Nuclear History Holds Lessons for the Future," *WMD Junction*, May 13, 2013, http://wmdjunction.com/130513_turkey_nuclear_history.htm.

9   "The Turkish Economy: Strong but Vulnerable," *Economist*, June 15, 2013, www. economist.com/news/europe/21579491-turkey-remains-highly-exposed-loss-confidence-foreign-investors-strong-vulnerable. Regarding investment-grade ratings, see Ye Xie and Selcuk Gokoluk, "Turkey Raised to Investment Grade by Moody's on Debt Cuts," Bloomberg News, May 17, 2013, www.bloomberg.com/news/2013-05-16/turkey-raised-to-investment-grade-by-moody-s-on-debt-cuts.html.

10 "The Turkish Economy: Strong but Vulnerable."

11 Mamta Badkar, "Turkey's Economic Mess in 5 Charts," *Business Insider*, March 23, 2014, www.businessinsider.com/whats-gone-wrong-in-turkey-2014-3.

12 Osman Bahadır Dinçer and Mustafa Kutlay, "Turkey's Power Capacity in the Middle East: Limits of the Possible," International Strategic Research Organization/ Uluslararası Stratejik Araştırmalar Kurumu (USAK), June 2012, 2.

13 Marc Champion, "Does Erdoğan Really Believe In an Interest-Rate Lobby?" *Bloomberg View*, June 28, 2013, www.bloomberg.com/news/2013-06-28/does-Erdoğan-really-believe-in-an-interest-rate-lobby-.html.

14 Tulin Daloglu, "Obama, Erdoğan Speak for the First Time Since Graft Probe," *Al-Monitor Turkey Pulse*, February 21, 2014, www.al-monitor.com/pulse/originals/2014/02/obama-Erdoğan-conversation-corruption-bribery-ideology.html.

15 Michael Koplow, "Erdoğan's Paranoia Hits an All-Time High," *Ottomans and Zionists* (blog), August 20, 2013, http://ottomansandzionists.com/2013/08/20/erdogans-paranoia-hits-an-all-time-high.

16 "Gül: Syria at Risk of Becoming Afghanistan on the Mediterranean," *Today's Zaman*, November 4, 2013, www.todayszaman.com/news-330594-gul-syria-at-risk-of-becoming-afghanistan-on-the-mediterranean.html.

17 İrem Karakaya and Bayram Kaya, "Syrian al-Qaeda Prepares to Launch Attack in Turkey's Big Cities," *Today's Zaman*, November 4, 2013, www.todayszaman.com/news-330595-syrian-al-qaeda-prepares-to-launch-attack-in-turkeys-big-cities.html.

18 Kadri Gursel, "Time Running Out for Turkey-PKK Peace Process," trans. Timur Goksel, *Al-Monitor*, November 4, 2013, www.al-monitor.com/pulse/originals/2013/11/akp-stall-kurd-peace-process.html.

19 See Varnum, "Turkey in Transition: Toward or Away From Nuclear Weapons?"

20 Koplow, "Erdoğan's Paranoia Hits an All-Time High."

21 *Public Opinion Surveys of Turkish Foreign Policy* 2013/3, Center for Economics and Foreign Policy Studies (EDAM), October 2013.

22 Ibid.

23 Richard Weitz, "Global Insights: Turkey's Russia Policy Put to the Test by Ukraine Crisis," *World Politics Review*, March 18, 2014, www.worldpoliticsreview.com/articles/13636/global-insights-turkey-s-russia-policy-put-to-the-test-by-ukraine-crisis.

24 Robert S. Norris and Hans M. Kristensen, "U.S. Tactical Nuclear Weapons in Europe, 2011," *Bulletin of the Atomic Scientists* 67, no. 1 (January/February 2011): 69.

25 Connie Pillich, "The Modern Costs of the Yesteryear Bomb," *Hill: Congress Blog*, June 13, 2013, http://thehill.com/blogs/congress-blog.

26 Varnum, "Turkey in Transition: Toward or Away From Nuclear Weapons?"

27 Interview with Dr. Joshua Walker, Director of Global Programs, APCO Worldwide, Washington, D.C., October 30, 2013.

28 Murat Karagöz, "U.S. Arms Embargo Against Turkey After 30 Years—An Institutional Approach Towards U.S. Policy Making," *Perceptions: Journal of International Affairs* (Winter 2004–2005): 107.

29 Burak Bekdil, "Turkey Distancing From Missile Deal With China," *Hürriyet Daily News*, March 11, 2014, www.hurriyetdailynews.com/turkey-distancing-from-missile-deal-with-china.aspx?pageID=238&nID=63415&NewsCatID=483.

30 Semih İdiz, "Turkey Faces 'Geography's Revenge' in Crimea," *Al-Monitor Turkey Pulse*, March 21, 2014, www.al-monitor.com/pulse/originals/2014/03/turkey-crimea-policy-russia-strategy-tatars-geography-rights.html.

31 "Turkey Calls for Elimination of WMDs From Middle East," *Today's Zaman,* November 1, 2013, www.todayszaman.com/news-330391-turkey-calls-for-elimination-of-wmds-from-middle-east.html.

32 Victor Gilinsky and Henry Sokolski, "The Iran Interim Agreement: An International Precedent for Nuclear Rules," *Bulletin of the Atomic Scientists*, December 6, 2013, http://thebulletin.org/iran-interim-agreement-international-precedent-nuclear-rules.

33 Aaron Stein, "Turkey Is in Trouble: Ankara's No-Win Syria Strategy," *Turkey Wonk: Nuclear and Political Musings in Turkey and Beyond* (blog), October 18, 2013, http://turkeywonk.wordpress.com/2013/10/18/turkey-is-in-trouble-ankaras-no-win-syria-strategy.

34 "UN Gave Strength to the Syrian Regime: Turkish PM," *Hürriyet Daily News*, September 2, 2013, www.hurriyetdailynews.com/un-gave-strength-to-the-syrian-regime-turkish-pm.aspx?pageID=238&nID=53655&NewsCatID=338.

35 Dan Bilefsky, "Turkey's Chief European Union Negotiator Acknowledges Turkey May Never Join Bloc," *New York Times, Lede* (blog) September 23, 2013, http://thelede.blogs.nytimes.com/2013/09/23/turkeys-chief-european-union-negotiator-acknowledges-turkey-may-never-join-bloc/?_r=0.

36 Ibid. See also "Turkish PM Erdoğan Accuses EU of 'Smear Campaign,'" *Hürriyet Daily News and Economic Review*, September 4, 2013, www.hurriyetdailynews.com/turkish-pm-Erdoğan-accuses-eu-of-smear-campaign.aspx?pageID=238&nID=53783&NewsCatID=338.

37 Emre Erşen, "The Shanghai Cooperation Organization: A New Alternative for Turkish Foreign Policy?" Middle East Institute, October 18, 2013, www.mei.edu/content/shanghai-cooperation-organization-new-alternative-turkish-foreign-policy.

38 Daniel Dombey, "Gul Hints at Political Ambitions in Speech to Turkey's Parliament," *Financial Times*, October 1, 2013, www.ft.com/intl/cms/s/0/e93a1908-2ab0-11e3-8fb8-00144feab7de.html#axzz2hddJPuMV.

39 See, for example, Şebnem Udum, "Turkey's Non-Nuclear Weapon Status—A Theoretical Assessment," presented at the 56th Pugwash Conference on Science and World Affairs: A Region in Transition: Peace and Reform in the Middle East, Cairo, Egypt 2006. Varnum, "Turkey in Transition: Toward or Away From Nuclear Weapons?"

40 See for example, Etel Solingen's domestic political survival model. Etel Solingen, *Nuclear Logics: Contrasting Paths in East Asia and the Middle East* (Princeton, N.J.: Princeton University Press, 2007).

41 Matthew Fuhrmann, "Spreading Temptation: Proliferation and Peaceful Nuclear Cooperation Agreements," *International Security* 34, no. 1 (Summer 2009).

42 For example, a frequently cited report for the U.S. Senate Foreign Relations Committee asserted: "… these three states in particular [Saudi Arabia, Egypt, and Turkey] appear to be moving deliberately in the direction of a nuclear hedging strategy that would position them to obtain a nuclear weapons breakout capability in the next two decades." "Chain Reaction: Avoiding a Nuclear Arms Race in the Middle East, Report to the Committee on Foreign Relations, United States Senate," February 2008, www.fas.org/irp/congress/2008_rpt/chain.pdf.

43 Jessica Varnum was the lead author of the Turkey Country Profile produced for the Nuclear Threat Initiative website. The profile details Turkey's extremely limited nuclear capabilities and facilities. "Turkey Country Profile: Nuclear Facilities," James Martin Center for Nonproliferation Studies, 2013, www.nti.org/country-profiles/turkey/facilities.

44 "Urgent Rethink Needed on Japan-Turkey Nuclear Energy Pact," *Asahi Shimbun* (editorial), January 8, 2014, http://ajw.asahi.com/article/views/editorial/AJ201401080039.

45 Jessica Varnum, "Overblown Rhetoric Exaggerates Proliferation Risks of Japan-Turkey Nuclear Cooperation," *New Atlanticist*, Atlantic Council, January 15, 2014, www.atlanticcouncil.org/blogs/new-atlanticist/overblown-rhetoric-exaggerates-proliferation-risks-of-japan-turkey-nuclear-cooperation.

46 "Turkish Energy Minister Denies Uranium Enrichment Intention," *Hürriyet Daily News*, January 9, 2014, www.hurriyetdailynews.com/turkish-energy-minister-denies-uranium-enrichment-intention.aspx?pageID=238&nID=60787&NewsCatID=348.

47 Jessica Varnum, "Closing the Nuclear Trapdoor in the U.S.-Turkey 'Model' Partnership: Opportunities for Civil Nuclear Cooperation," Turkey Project Policy Paper, Center on the United States and Europe, Brookings Institution, June 17, 2013.

48 While producing gravity bombs would enable Turkey to get around the need for a strategic missile delivery system (including missile development and warhead miniaturization), an arsenal that relies exclusively on strategic bombers to deliver nuclear weapons is problematic given modern air defense systems.

49 Henri J. Barkey, "Turkey's Perspectives on Nuclear Weapons and Disarmament," in *Unblocking the Road to Zero: Perspectives of Advanced Nuclear Nations*, ed. Barry Blechman (Washington, D.C.: Stimson Center, 2009), 73.

50 Previously cited in Varnum, "Turkey in Transition: Toward or Away From Nuclear Weapons?" "Middle East Nations Reportedly Seeking Atomic Weapons Capability," Open Source Center EUP20070416338011; Vincent Jauvert, "The Doctor Strangelove of the Middle East," *Le Nouvel Observateur* (French), April 12–18, 2007, 64–67.

51 Scott Sagan, "Chapter 5: Nuclear Latency and Nuclear Proliferation," Potter and Mukhatzhanova, eds., *Forecasting Nuclear Proliferation in the 21st Century: Volume 1, The Role of Theory* (Stanford, Calif.: Stanford University Press, 2010), 81.

52 Işık Özel, "The Politics of De-Delegation: Regulatory (In)Dependence in Turkey," *Regulation and Governance* 6, no. 1 (March 2012): 120.

53 Varnum, "Closing the Nuclear Trapdoor in the U.S.-Turkey 'Model' Partnership: Opportunities for Civil Nuclear Cooperation."

54 As further discussed in the author's Brookings study, Turkey's regulatory and operating challenges as a nuclear newcomer will be greatly exacerbated by the fact that "… neither the VVER-1200 nor the Atmea-1 is in operation anywhere, and anyone qualified to evaluate either design likely has a conflict-of-interest relationship with the vendor." See ibid.

55 Aaron Stein, "2025: Turkey's Looming Nuclear Weapons Deadline," *Turkey Wonk: Nuclear and Political Musings in Turkey and Beyond* (blog), November 19, 2012, http://turkeywonk.wordpress.com/2012/11/19/2025-turkeys-looming-nuclear-weapons-deadline.

56 Hans M. Kristensen, "B61–12 Nuclear Bomb Integration on NATO Aircraft to Start in 2015," FAS *Strategic Security* (blog), March 13, 2014, http://blogs.fas.org/security/2014/03/b61-12integration.

57 Varnum, "Turkey in Transition: Toward or Away From Nuclear Weapons?"

58 It is, of course, conceivable that Turkey could illicitly purchase the requisite fissile materials, but even if it could find a willing seller, this is a more appropriate route to nuclear weapons for a non-state actor, as states have generally taken the approach that a nuclear deterrent requires the capability to produce more than one or two nuclear weapons.

59 Pakistani Prime Minister Zulfikar Ali Bhutto famously asserted in 1974 that "If India builds the bomb, we will eat grass or leaves, even go hungry, but we will get one of our own." Pakistan and North Korea were often dismissed by analysts as countries incapable of acquiring nuclear weapons, but both proved that even the most impoverished countries can do so if they are willing to exchange overall national welfare for nuclear weapons. Feroz Hassan Khan, *Eating Grass: The Making of the Pakistani Bomb* (Stanford, Calif.: Stanford University Press, 2012).

60 Soner Cagaptay, et al., "Turkey's Transformation: Prospects and Limits," Washington Institute, March 7, 2014, www.washingtoninstitute.org/policy-analysis/view/turkeys-transformation-prospects-and-limits.

61 Varnum, "Turkey in Transition: Toward or Away From Nuclear Weapons?"

CONCLUSION

GEORGE PERKOVICH

T urkey is a rising economic, political, and strategic actor in the international system. Notwithstanding recent internal turmoil and unfolding crises in states next to it, Turkey will continue to attract international interest centered on its own political economy and on its capacity to affect dynamics in the greater Middle East, the Caucasus, and Central Asia. Turkey also will remain an important actor in the international nuclear order. Its size and economic growth make it a potentially significant market for nuclear energy technology, whether supplied by international partners or developed indigenously. And Turkey's location in a tumultuous region naturally leads to speculation that Turkish leaders could consider seeking options to develop nuclear weapons.

Notwithstanding Turkey's importance in all of these contexts, remarkably little informed analysis and debate on its nuclear future are taking place in Turkey or broader international society. Prospects for building and operating nuclear power plants tend to be asserted with scant rigorous assessment of feasibility, just as Turkey's potential interest in acquiring nuclear weapons is sometimes asserted with little consideration of the feasibility and costs and benefits of attempting such an acquisition.

The authors and editors hope this volume will stimulate further efforts to add factual richness, historical perspective, analytical rigor, and strategic insight to discussions of Turkey's nuclear future. Among the major contributions these chapters make, several stand out.

Turkey has been interested in producing nuclear electricity for decades and has sound economic reasons to be interested in non-fossil-fuel sources of power, especially in view of the paucity of its primary energy resources and the growing domestic demand for electricity. In other words, Turkey's aspiration to switch to nuclear power is understandable within the context of this emerging country's energy needs. Proliferation aims do not have to be imputed to explain the scale and aims of this program.

It is sometimes asserted that foreign pressure has been largely responsible for blocking Turkey's efforts to develop nuclear energy. However, as Gürkan Kumbaroğlu demonstrates in chapter 1, domestic factors within Turkey caused plans for nuclear power to be abandoned from the 1970s through 2010. In the late 1970s, Sweden and Turkey could not agree on financing terms, and then the 1980 coup in Turkey created political uncertainty in the nuclear domain. In 1983, Turkey invited nuclear power plant vendors from Germany, Canada, and the United States to submit proposals for building nuclear power reactors. The U.S. company, General Electric, abandoned its effort due to concerns about seismic instability at the site Turkey had selected. Then Siemens, the German company, removed itself from negotiations due to disagreements over the financing and partnership arrangements Turkey sought. Finally, in 1987, Turkey and the Canadian firm, Atomic Energy of Canada Limited, proved unable to conclude an agreement because neither side was willing to provide extensive financial guarantees, and Turkey would not meet the company's requests for protection against risk.

Similar concerns over financing and risk management derailed negotiations in the mid-1990s between Turkey and two international consortia, one involving Siemens and the French company Framatome, and the second involving the U.S. company Westinghouse and Japan's Mitsubishi. In 2000, the government of Turkey, beset by a financial crisis, postponed consideration of building nuclear power plants. Finally, in 2010, the Russian Federation made an exceptionally attractive offer to Turkey to finance and build, operate, and own nuclear power reactors at the Akkuyu site in Turkey. Turkey remains interested in working with other vendors to build power reactors but has struggled to find others willing to meet Turkey's terms. In 2013, the French-Japanese team of Areva-GDF Suez and Mitsubishi Heavy Industries reached agreement with Turkey to build four reactors, with construction nominally to begin in 2017.

The Turkish experience recorded by Gürkan Kumbaroğlu illustrates the fact that, despite claims by sellers of nuclear power plants and scientists and engineers who wish to lead nuclear programs, it is extremely expensive to finance, design, and construct nuclear power plants. Russia's offer to Turkey was exceptionally generous in many ways, which is proved by the reluctance of other vendors to offer similar terms.

Beyond financing, it is also quite difficult to build and operate nuclear plants with the levels of safety that citizens rightfully demand. The design expertise, equipment, and technology required are developed slowly and at considerable expense, which is one reason that new entrants into the nuclear field tend to rely initially on supplies from established vendors. Beyond finances and technical infrastructure, a safe and cost-effective nuclear industry requires complicated social and institutional infrastructure in the form of laws, independent regulatory authorities, trained operators and regulators, emergency management programs, a sound safety culture, and so on.

As chapter 2 by İzak Atiyas describes, Turkey is in the midst of creating the legal and operational basis for a robust, independent regulatory agency that will give Turks and the international community confidence that safety will be an overriding imperative. The process and prospect of establishing an adequate regulatory framework and agency will depend on the fundamental evolution of the Turkish state. Atiyas rightly underlines current threats to the independence of regulatory institutions and more generally the lack of a consensus about the role of independent institutions in policymaking. Atiyas succinctly explains how "the current regulatory authority, and the regulatory framework in general [in Turkey], does not yet satisfy the requirements of independence, transparency, and accountability" that international guidelines deem as necessary to safely build and operate nuclear power plants. Turkey is not alone in this regard, as similar challenges face, or have faced, Japan, India, South Korea, Russia, and the United States, to name a few.

Nor is Turkey unique in fostering confusion at home and abroad about its nuclear plans. Turkish authorities have not persuasively developed and then shared with the Turkish public their strategic outlook for nuclear energy. The lack of a proper and informed debate within the country about the need for nuclear power—and more importantly about the planned safety and security measures—tends to fuel the polarization around Ankara's decision to switch to nuclear energy.

In toto, the early chapters in this volume clarify the challenges and uncertainties surrounding the future of nuclear energy in Turkey, filling an information gap in most discussions of nuclear energy in the country. These chapters add detail and perspective that could be valuable for Turks who are interested in serious deliberations on whether and how the country can best pursue nuclear sources of energy.

Much as discussions of nuclear energy often are infused with exaggerated assertions of benefits and risks, so, too, international discussions of Turkey's potential acquisition of nuclear weapons often are oversimplified. The chapters in this volume offer corrections in several dimensions.

There is a chicken-and-egg riddle in analyzing how nuclear weapon programs are "born." Do states determine that it is feasible for them to acquire nuclear weapons before they decide that it is desirable to do so, or do leaders decide to "go for the bomb" and then let their technicians try to solve the feasibility problems? History shows that sometimes leaders have decided they definitely want their technicians to produce nuclear weapons and then confront the challenges of feasibility. This occurred most obviously in the United States, the Soviet Union, and Pakistan. In other cases, leaders have not been sure whether they actually wanted to acquire nuclear weapons, and first tasked personnel to explore feasibility. This approximately describes what happened in India and, thus far, Iran.

The chapters in this volume indicate that Turkish leaders to date have not authorized extensive explorations of either the feasibility or the desirability of acquiring an independent nuclear arsenal. On the feasibility side of the equation, as Mustafa Kibaroğlu and Jessica Varnum describe in chapters 7 and 8, respectively, publicly available information indicates that Turkey has not sought and does not possess the equipment, material, design information, and multifaceted expertise necessary to produce a usable nuclear arsenal. The range of capabilities required is enormous and difficult to develop. It entails, among other things: the production of highly enriched uranium or plutonium; equipment, facilities, and know-how to form highly enriched uranium or plutonium into nuclear-weapon cores; the design and components required for the non-nuclear package that makes a weapon detonate; engineering and testing of neutron initiators to ensure efficient chain reaction; delivery systems and targeting systems; and training of units to manage and deliver nuclear weapons. As Mark Hibbs (chapter 6) and other authors explain, Turkey, as a non-nuclear-weapon state under the Treaty on the Non-Proliferation of Nuclear Weapons (NPT), is obligated to undertake only peaceful nuclear activities. Efforts to acquire capabilities to make and deploy nuclear weapons would violate this commitment. Thus, if Turkey were detected to be seeking such capabilities, it would face enormous international pressures and costs, including the probable disruption of its now-underway efforts to reap the benefits of nuclear electricity. These and other factors would greatly complicate the feasibility of Turkey's acquiring nuclear weapons.

The development of ballistic missile and missile defense capabilities could serve conventional military purposes as well as nuclear deterrence functions. Turkey's membership in the NPT and other nuclear non-proliferation arrangements does not preclude development of missiles or missile defenses. As Aaron Stein reports in chapter 5, Turkey's ambitious missile-related programs could enable it to enhance its deterrence capability whether or not Ankara also sought to acquire nuclear weapons.

Regarding desirability, the chapters by Doruk Ergun (chapter 3), Kibaroğlu, and Varnum detail a number of factors that weigh against Turkey's seeking its own nuclear weapons. Turkey continues to benefit from its membership in the North Atlantic Treaty Organization (NATO), albeit with occasional frustrations and doubts. As a country hosting U.S. and NATO nuclear weapons, Turkey stands under the center of the NATO nuclear deterrence umbrella. This arrangement provides Turkey with a degree of nuclear deterrence that is much less risky and costly than would exist if and when the country undertook a time-consuming and fraught campaign to acquire nuclear weapons of its own. To be sure, Turkey's allies in NATO share a responsibility to uphold Ankara's confidence in extended deterrence, which may not always be appreciated by political parties and states in the safer Western environs of the alliance. In any case, the most pressing security threats that loom on Turkey's borders today and on the horizon tomorrow would not be deterrable or defeatable by nuclear weapons. Meanwhile, a Turkish quest to acquire an independent nuclear arsenal would create new threats of sanction, isolation, and countermeasures by neighboring states.

Thus, in terms of desirability and feasibility, the analyses in this volume conclude that for the foreseeable future Turkish leaders will determine that the national interest is best served by eschewing moves to acquire an independent nuclear force. The only scenario in which Turkey could contemplate seeking its own nuclear deterrent would be in the unlikely case the security relationship with NATO and the United States would collapse. Even an end to Turkey's EU ambitions is not seen as a critical factor likely to affect Turkish policymakers' approach to nuclear deterrence. To put a finer point on this conclusion, Turkey appears now not to have an interest even in ambiguously hedging its position by pushing toward the boundaries that distinguish a purely peaceful nuclear program under the NPT from a militarily oriented nuclear program. One obvious boundary is the indigenous production of fissile materials—particularly enriched uranium and separated plutonium. While Turkey has insisted that it and other states have a sovereign right to engage in fuel-cycle activities, Ankara's approach to nuclear generation of electricity remains focused on importing plants and fuel from well-established international suppliers, which is the most economical and technically feasible way for a country in Turkey's position to proceed.

This book also covers Turkey's role in international nuclear governance and the nonproliferation regime. As a country with aspirations to acquire a more influential role in shaping global governance, Turkey has been increasingly active in international fora responsible for managing the nuclear world. Hibbs relates Ankara's efforts within the Nuclear Suppliers Group to maintain the balance between nonproliferation commitments and assistance to countries interested in transitioning to nuclear power embedded in the NPT. In this regard, Turkey has diplomatically contested efforts by the United States, the European Union (EU) member states, and other industrialized nations to focus on restricting the transfer of technology. Occasionally, as a result, Turkey's diplomatic priorities as a key regional player have clashed with its willingness to deepen its collaboration with its Western partners in the field of nonproliferation. But overall Turkey remains strongly committed to strengthening the non-proliferation regime.

If the foregoing are some of the most important factual and analytic contributions that this book's authors have made, they help identify key questions that Turkey's leaders and others may wish to explore in the coming years.

How might the fundamental tensions currently seen in Turkey regarding the distribution of power within the state be resolved, and will this resolution affect the processes and interests that would shape future nuclear policymaking? Similarly, are there risks to Turkey's Western orientation, and if so, what would this entail for Turkey's nuclear ambitions?

How will the construction, operation, and regulation of the nuclear power reactors now being constructed by Russia turn out? How will these results affect Turkey's interest in nuclear energy and the means by which further nuclear electricity generation would be pursued? How will other potential nuclear power plant exporters view these developments?

Will the shale gas "revolution" and other electricity-production developments affect Turkey's calculations of the desirability of nuclear energy, and if so, how?

How will the external environment affecting Turkey's security evolve in the coming years? Will Turkey's diplomatic and security policies and related intelligence, military, and diplomatic capabilities be adapted to keep pace with possible changes in this environment?

Will NATO maintain, augment, or weaken confidence that the alliance will defend Turkey's security interests if and when these interests are challenged by external actors? Specifically—but not necessarily most important—will NATO manifest resolve and capabilities to extend nuclear deterrence on Turkey's behalf in case of severe threats?

Will the challenge posed by Iran's heretofore ambiguous nuclear program be resolved diplomatically in ways that alleviate concerns by Turkey and other states that Iran might acquire nuclear weapons? Or, if Israel or the United States or both take military action to stop (or slow) an Iranian effort to acquire nuclear weapons, how will the ensuing developments affect Turkey's interests?

These are big questions. The answers to most of them will emerge over the next decade and more. The future of nuclear energy and nuclear weapons—the nuclear order—in the twenty-first century is uncharted today. Turkey is located at the juncture of Central Asia, the Middle East, and Europe. As such it will be an increasingly important shaper of the regional and international nuclear order. If this volume serves as a useful baseline for policymakers, experts, and journalists to explore these questions and the direction of the nuclear order into the future, it will have achieved its purpose.

INDEX

Note: Tables, figures, and notes are indicated by t, f, and n.

CONTRIBUTORS

İzak Atiyas is a professor of European studies and coordinator of the MA program in Public Policy at Sabancı University, Istanbul. His research areas include productivity, industrial policy, competition policy, regulation of network industries, and privatization. Atiyas was previously a senior economist in the World Bank's Private Sector Development Department (1988–1995) and a visiting assistant professor of economics at Bilkent University (1995–1998). Since January 2011 he is also the director of the TUSİAD-Sabancı University Competitiveness Forum, which undertakes research on Turkey's international competitiveness. Atiyas received his BA from Boğaziçi University's department of economics in 1982, and his PhD in economics from New York University in 1988.

Doruk Ergun is a research fellow at the Center for Economics and Foreign Policy Studies (EDAM), where he works on Turkish foreign policy and security issues. He was previously a research assistant at the NATO Parliamentary Assembly. Ergun received his MA in international affairs with a focus on international security studies from the George Washington University in 2011 and his BA in social and political sciences from Sabancı University in 2009.

Mark Hibbs is a senior associate in Carnegie's Nuclear Policy Program, based in Berlin. Before joining Carnegie, he was an editor and correspondent for nuclear energy publications including *Nucleonics Week* and *Nuclear Fuel*. From the late 1980s until the mid-1990s, he covered nuclear developments in the Soviet bloc, including research on the USSR's nuclear fuel-cycle facilities and its nuclear materials inventories. Since the mid-1990s, his work has focused on emerging nuclear programs in Asia, including China and India. Throughout the last two decades, many of the over 3,000 articles he wrote investigated nuclear proliferation–related developments in Argentina, Brazil, China, India, Iran, Iraq, Israel, Japan, Libya, North and South Korea, Pakistan, South Africa, Syria, and Taiwan. Since 2003, he has made many detailed findings about clandestine procurement in Europe related to gas centrifuge uranium enrichment programs in Iran, Libya, North Korea, and Pakistan.

Can Kasapoğlu is a security studies specialist and military analyst. He has held several visiting researcher posts, including at the Begin-Sadat Center for Strategic Studies (BESA Center) in Israel (2012) and at the Fondation pour la Recherche Stratégique in France (2014). He is currently a research fellow in the Istanbul-based Center for Economics and Foreign Policy Studies and a faculty member at Girne American University. Kasapoğlu specializes in war studies, strategic weapon systems, missile defense, biological and chemical warfare, low-intensity conflicts, terrorism, strategic intelligence, and civil-military relations. In addition, he also focuses on strategic affairs in the Middle East, Iranian military modernization, and Turkish-Israeli relations. Kasapoğlu holds a PhD from the Strategic Research Institute at the Turkish War College, and an MSci degree from the Defense Sciences Institute at the Turkish Military Academy.

Mustafa Kibaroğlu is the chair of the department of political science and international relations at the newly established MEF University in Istanbul. He previously served as chair of the department of international relations at Okan University, Istanbul, and as vice chair of the department of international relations at Bilkent University, Ankara. His research centers on the proliferation of weapons of mass destruction, Middle Eastern politics, and Turkish foreign policy. Kibaroğlu has held fellowships at the Belfer Center for Science and International Affairs, Harvard University; the Center for Nonproliferation Studies, the Monterey Institute of International Studies; and the United Nations Institute for Disarmament Research in Geneva. He is the author (with Ayşegül Kibaroğlu) of *Global Security Watch—Turkey: A Reference Handbook* (Praeger Security International, 2009) and has edited and contributed to several edited volumes on Turkish security issues. He received his PhD in international relations from Bilkent University in 1996.

Gürkan Kumbaroğlu is a professor of industrial engineering at Boğaziçi University, Istanbul. He is the recipient of several awards, including Turkey's first Energy Oscar Award in the category of Scientific Contribution to Energy Markets (2011) and the Visiting Professorship Award for Senior International Scientists, given by the Chinese Academy of Sciences (2012). Kumbaroğlu is an editorial board member of the international journals *Sustainability, Innovative Energy Policies*, and *Journal of Self-Governance and Management Economics*. He serves on various scientific boards related to energy and the environment and has published numerous articles in peer-reviewed journals as well as book chapters on energy and environmental policy. He is the founding president of the Turkish Association for Energy Economics and 2015 president-elect of the International Association for Energy Economics. He is also a member of the executive and supervisory board of the Center for Economics and Foreign Policy Studies.

George Perkovich is vice president for studies at the Carnegie Endowment for International Peace. His research focuses on nuclear strategy and nonproliferation, with a concentration on South Asia, Iran, and the problem of justice in the international political economy. Perkovich is author of the award-winning book *India's Nuclear Bomb* (University of California Press, 2001) and co-author of the Adelphi Paper "Abolishing Nuclear Weapons" (2008). This paper is the basis of the book *Abolishing Nuclear Weapons: A Debate*, which includes seventeen critiques by thirteen eminent international commentators. Perkovich served as a speechwriter and foreign policy adviser to Senator Joe Biden from 1989 to 1990 and is an adviser to the International Commission on Nuclear Nonproliferation and Disarmament and member of the Council on Foreign Relations' task force on U.S. nuclear policy.

Aaron Stein is an associate fellow at the Royal United Services Institute (RUSI). He is also a researcher at the Center for Economics and Foreign Policy Studies in Istanbul, where he works on security and proliferation issues in the Middle East. He is currently a PhD candidate at King's College London, researching Iranian and Turkish nuclear decisionmaking. Stein has written extensively on Turkish politics and regional proliferation, publishing in scholarly journals and print media, including the *New York Times*, *Foreign Affairs*, *Foreign Policy*, the *Bulletin of the Atomic Scientists*, the *National Interest*, and *World Politics Review*. He holds a BA in political science from the University of San Francisco and an MA in international policy studies with a specialization in nuclear nonproliferation from the Monterey Institute of International Studies.

Sinan Ülgen is a visiting scholar at Carnegie Europe in Brussels, where his research focuses on the implications of Turkish foreign policy for Europe and the United States, nuclear policy, and the security and economic aspects of transatlantic relations. He is a founding partner of Istanbul Economics, a Turkish consulting firm that specializes in public and regulatory affairs, and chairman of the Center for Economics and Foreign Policy Studies, an independent think tank in Istanbul. Ülgen has served in the Turkish Foreign Service in several capacities: in Ankara at the United Nations desk (1990–1992); in Brussels at the Turkish Permanent Delegation to the European Union (1992–1996); and at the Turkish embassy in Tripoli (1996). He is a regular contributor to Turkish dailies, and his opinion pieces have been published in major newspapers. He is the author of *The European Transformation of Modern Turkey* with Kemal Derviş (Center for European Policy Studies, 2004) and *Handbook of EU Negotiations* (Bilgi University Press, 2005).

Jessica C. Varnum is the Nuclear Threat Initiative project manager at the Center for Nonproliferation Studies and an adjunct professor at the Monterey Institute of International Studies. Varnum is an expert on U.S.-Turkey relations, focusing on the strategic dimensions of the relationship, the effects of evolving Turkish domestic politics on the bilateral alliance, and the role of the North Atlantic Treaty Organization. She regularly lectures, writes, and contributes to research and strategic dialogue projects aimed at understanding and improving bilateral relations. She is the author of "Turkey in Transition: Toward or Away from Nuclear Weapons?" in *Forecasting Nuclear Proliferation in the 21st Century: A Comparative Perspective* (Stanford University Press, 2010), and a 2013 occasional paper for the Brookings Institution on Turkey's nuclear power program. Her work has appeared in *Nonproliferation Review, World Politics Review,* and the *International Herald Tribune,* and on the Nuclear Threat Initiative website. Varnum earned an MA in International Policy Studies with a certificate in nonproliferation studies from the Monterey Institute of International Studies, a BA in government and international studies, summa cum laude, from Colby College, and is pursuing a PhD in U.S.-Turkey relations with King's College London's Defense Studies Department.

CARNEGIE ENDOWMENT FOR INTERNATIONAL PEACE

The Carnegie Endowment for International Peace is a unique global network of policy research centers in Russia, China, Europe, the Middle East, and the United States. Our mission, dating back more than a century, is to advance the cause of peace through analysis and development of fresh policy ideas and direct engagement and collaboration with decisionmakers in government, business, and civil society. Working together, our centers bring the inestimable benefit of multiple national viewpoints to bilateral, regional, and global issues.